数学专业英语

English for Mathematics

主编 郝翠霞
副主编 程美玉 马维军

哈尔滨工业大学出版社

内 容 提 要

本书为各高等院校数学及相关专业英语基础教材,也适用于从事数学方面理论研究的读者参考。本书共设七个单元,主要内容涉及:数学史、数学分析、高等代数、概率论等学科中的基础经典问题以及具有代表性的内容。另外,本书的词汇表中列出了国家已经确定的和被广泛认同的专业术语,便于读者了解和查阅。

图书在版编目(CIP)数据

数学专业英语/郝翠霞主编.—哈尔滨:哈尔滨工业大学出版社,2005.2(2021.7重印)
ISBN 978-7-5603-2118-9

Ⅰ.数… Ⅱ.郝… Ⅲ.数学-英语-高等学校-教材 Ⅳ.H31

中国版本图书馆 CIP 数据核字(2005)第 013684 号

策划编辑	段琛妹 杨 桦
封面设计	卞秉利
出版发行	哈尔滨工业大学出版社
社　　址	哈尔滨市南岗区复华四道街 10 号　邮编 150006
传　　真	0451-86414749
网　　址	http://hitpress.hit.edu.cn
印　　刷	哈尔滨市圣铂印刷有限公司
开　　本	880mm×1230mm　1/32　印张 6.5　字数 168 千字
版　　次	2005 年 2 月第 1 版　2021 年 7 月第 7 次印刷
书　　号	ISBN 978-7-5603-2118-9
定　　价	20.00 元

(如因印装质量问题影响阅读,我社负责调换)

Preface
Hao Cuixia 2006 – 6

With China's joining the WTO and the rapid development of science and technology in modern society, English is playing an important role in more and more fields. The international exchange in distinct disciplines is increasing every day. Both mathematical knowledge and foreign language ability are required of mathematician in our times. Teaching of English for mathematics thus becomes essential. This book is compiled to meet the requirements of the students in mathematics department. It is based on College English Curriculum(CEC). According to CEC, the aim of college English teaching is to cultivate students to gain strong ability of reading, a fair ability of listening & translating and the basic mastery of skills of writing & speaking, so that they are able to use English as a tool to obtain new information in their field. This is also expected to lay a foundation for the further improvement of their English ability. The book covers topics of several main branches of mathematics. Grateful acknowledgement is presented here to the staff of School of Mathematical Science of Heilongjiang University for their great support in the compiling process of this book.

Since the writing work was done in a rush, errors and mistakes could hardly be avoided. Suggestions or comments are therefore very welcome.

Some edition material is cited from π *in the Sky* etc.. If something inappropriate, please point out so that they can be corrected.

If there is a concern for the issue of royalty in books, contact us.

CONTENTS

1 Math Story and Strategies
- 1.1 THE NUMBER π 1
- 1.2 THE NUMBER e 5
- 1.3 INEQUALITIES FOR CONVEX FUNCTION 11

2 The History
- 2.1 THE INTEGRAL 24
- 2.2 THE DIFFERENTIAL 33
- 2.3 THE CALCULUS 42
- 2.4 SHORT BIOGRAPHIES 50

3 The Algebra
- 3.1 NOTATIONS AND A REVIEW OF NUMBERS 60
- 3.2 LINEAR ALGEBRA 68
- 3.3 MATRIX ALGEBRA 83
- 3.4 EIGENVALUES AND EIGENVECTORS 86

4 Probability
- 4.1 INTRODUCTION TO PROBABILITY 94

5 Selected Problems
- 5.1 THE REAL NUMBER SYSTEMS 103
- 5.2 SELECTED PROBLEMS FROM THE INFINITE SERIES 106
- 5.3 DIFFERENTIAL EQUATION 113
- 5.4 FERMAT'S LAST THEOREM 117

5.5 REVEALING GOLDBACH CONJECTURE 122

6 Others

6.1 MATHEMATICS IN TODAY'S FINANCIAL MARKETS 129

6.2 MATHS AND MOTHS 139

6.3 RECKONING AND REASONING 154

6.4 AM I REALLY SICK? 161

7 Vector Space Some Problems Over Skewfield

7.1 BASIC CONCEPT 167

7.2 PROPERTY OF TRANSVECTION 173

7.3 QUASICENTER MATRIX 177

Vocabulary Index 181

Math Story and Strategies

1.1 THE NUMBER π

The most famous quantity in mathematics is the ratio of the circumference of a circle to its diameter, which is also known as the number pi and denoted by the Greek letter π

$$\frac{\text{circumference}}{\text{diameter}} = \pi$$

The symbol π was not introduced until just over two hundred years ago. The ancient Babylonians estimated this ratio as 3 and, for their purposes, this approximation was quite sufficient. According to the Bible, the ancient Jews used the same value of π. The earliest known trace of an approximate value of π was found in the Ahmes Papyrus written in about 16th century B.C., in which, indirectly, the number π is referred to as equal to 3.160 5. Greek philosopher and mathematician Archimedes, who lived about 225 B.C., estimated the value of pi to be less than $3\frac{1}{7}$ but more than $3\frac{10}{71}$. Ptolemy of Alexandria (c. 150 B.C.) gives the value of π to be about 3.141 6. In the far East, around 500

A.D., a Hindu mathematician named Aryabhata, who worked out a table of a sines, used for π the value 3.141 6. Tsu Chung-Chih of China, who lived around 470, obtained that π has a value between 3.141 592 6 and 3.141 592 7, and after him no closer calculation of π was made for one thousand years. The Arab Al Kashi about 1 430 obtained the amazingly exact estimated value for π of 3.141 592 653 589 793 2. There were several attempts made by various mathematicians to compute the value of π to 140, then 200, then 500 decimal places. In 1853, William Shanks carried the value of π to 707 decimal places. However, nobody seemed to be able to give the exact value for the number π.

What is the exact value of the number π? A mathematician made an experiment in order to find his own estimation of the number π. In his experiment, he used an old bicycle wheel of diameter 63.7 cm. He marked the point on the tire where the wheel was touching the ground and he rolled the wheel straight ahead by turning it 20 times. Next, he measured the distance traveled by the wheel, which was 39.69 meters. He divided the number 3 969 by 20×63.7 and obtained 3.115 384 615 as an approximation of the number π. Of course, this was just his estimate of the number π and he was aware that it was not very accurate.

The problem of finding the exact value of the number π inspired scientists and mathematicians for many centuries before it was solved in 1 761 by Johann Heinrich Lambert(1728 ~ 1777). Lambert proved that the number π cannot be expressed as a fraction or written in a decimal form using only a finite number of digits. Any such representation would always be only an estimation of the number π. Today, we call such numbers *irrational*. The ancient Greeks already knew about the existence of irrational numbers, which they called *incommensurables*. For example, they knew that the length of the diagonal of a square, with side of length equal to one length unit, is such a value. This value, which is denoted $\sqrt{2}$ and is equal to the number x such that $x^2 = 2$, cannot be expressed as a fraction.

Today in schools we use the estimation 3.14 for the number π, and of course this is completely sufficient for the type of problems we discuss in class. However, it was quickly noticed that in real life we need a better estimate to find more accurate measurements for carrying out construction projects, sea navigations and military applications. For most practical purposes, no more than 10 digits of π are required. For mathematical computation, even with astronomically precise calculations, no more than fifty exact digits of π are really necessary: 3.1415926535 8979323846 2643383279 5028841971 6939937510. However, with the power of today's supercomputers, we are able to compute more than hundreds of billions of digits of the number π. You can download at the web site http://www.verbose.net
files with the exact digits of the number π up to 200 million decimals. We also have the following approximations of the number π:

3.1415926535 8979323846 2643383279 5028841971 6939937510
   5820974944 5923078164 0628620890 8628034825 3421170679
   8214808651 3282306647 0938446095 5058223172 5359408128
   4811174502 8410270193 8521105559 6446229489 5493038196
   4428810975 6659334461 2847564823 3786783165 2712019091
   4564856692 3460348610 4543266482 1339360726 0249141273
   7245870066 0631558817 4881520920 9628292540 9171536436
   7892590360 0113305305 4882046652 1384146951 9415116094
   3305727036 5759591953 0921861173 8193261179 3105118548
   0744623799 6274956735 1885752724 8912279381 8301194912

Since the number π is the ratio of the circumference of a circle to its diameter, we can write a formula for the circumference of a circle, which is

$$C = \pi d$$

where C denotes the circumference and d denotes the diameter of the circle. If r denotes the radius of the circle then $d = 2r$, and we can rewrite the formula for the circumference as

$$C = 2\pi r$$

Words and Expressions

quantity *n.* 量,数量
ratio *n.* 比,比率
circumference *n.* 周长,圆周
circle *n.* 圆,圈,圆周
the circumference of a circle 圆的周长
diameter *n.* 直径
the ratio of the circumference of a circle to its diameter 圆周长与其直径的比
denote *vt.* 指示,表示
be denoted by 被表示为
approximate *adj.* 逼近的,近似的
approximation *n.* 逼近,近似;近似法;近似值
approximate value 近似值
Archimedes 阿基米德
pi π(圆周率)
divide *vt.* 除,等分
fraction *n.* 分数,小数;分式
decimal *adj.* 小数的;十进制的 *n.* 小数;十进小数
decimal form 小数形式
irrational *adj.* 无理的 *n.* 无理数
irrational number 无理数
length *n.* 长,长度
diagonal *n.* 对角线 *adj.* 对角线的
square *n.* 方,正方形 *vt.* 平方,二次幂
length of the diagonal of a square 正方形对角线的长度
unit *n.* 单位,单元

one length unit 一个长度单位
digit $n.$ 数字
compute $vt.$ 计算
formula $n.$ 公式
radius $n.$ 半径

1.2　THE NUMBER *e*

　　One of the first articles which we included in the "History Topics" section archive was on the history of π. It is a very popular article and has prompted many to ask for a similar article about the number *e*. There is a great contrast between the historical developments of these two numbers and in many ways writing a history of *e* is a much harder task than writing one for π. The number *e* is, compared to π, a relative newcomer on the mathematical scene.

　　The number *e* first comes into mathematics in a very minor way. This was in 1618 when, in an appendix to Napier's work on logarithms, a table appeared giving the natural logarithms of various numbers. However, that these were logarithms to base *e* was not recognized since the base to which logarithms are computed did not arise in the way that logarithms were thought about at this time. Although we now think of logarithms as the exponents to which one must raise the base to get the required number, this is a modern way of thinking. We will come back to this point later in this essay. This table in the appendix, although carrying no author's name, was almost certainly written by Oughtred. A few years later, in 1624, again *e* almost made it into the mathematical literature, but not quite. In that year Briggs gave a numerical approximation to the base 10 logarithm of *e* but did not mention *e* itself in his work.

　　The next possible occurrence of *e* is again dubious. In 1647 Saint Vincent computed the area under a rectangular hyperbola. Whether he recognised the connection with logarithms is open to debate, and even if he did there was little reason for him to come across the number e explicitly. Certainly by 1661 Huygens understood the relation between the rectangular hyperbola and the logarithm. He examined explicitly the relation between the area under the rectangular hyperbola $yx = 1$ and the logarithm. Of course, the number e is such that the area under the rectangular hyperbola from 1 to e is equal to 1. This is the property that makes e the base of natural logarithms, but this was not understood by mathematicians at this time, although they were slowly approaching such an understanding.

Huygens made another advance in 1661. He defined a curve which he calls "logarithmic" but in our terminology we would refer to it as an exponential curve, having the form $y = ka^x$. Again out of this comes the logarithm to base of e, which Huygens calculated to 17 decimal places. However, it appears as the calculation of a constant in his work and is not recognized as the logarithm of a number (so again it is a close call but e remains unrecognized).

Further work on logarithms followed which still does not see the number e appear as such, but the work does contribute to the development of logarithms. In 1668 Nicolaus Mercator published *Logarithmotechnia* which contains the series expansion of $\log(1 + x)$. In this work Mercator uses the term "natural logarithm" for the first time for logarithms to base e. The number e itself again fails to appear as such and again remains elusively just round the corner.

Perhaps surprisingly, since this work on logarithms had come so close to recognizing the number e, when e is first "discovered" it is not through the notion of logarithm at all but rather through a study of compound interest. In 1683 Jacob Bernoulli looked at the problem of compound interest and, in examining continuous compound interest, he tried to find the limit of $\left(1 + \dfrac{1}{n}\right)^n$ as n tends to infinity. He used the binomial theorem to show that the limit had to lie between 2 and 3 so we could consider this to be the first approximation found to e. Also if we accept this as a definition of e, it is the first time that a number was defined by a limiting process. He certainly did not recognise any connection between his work and that on logarithms.

We mentioned above that logarithms were not thought of in the early years of their development as having any connection with exponents. Of course from the equation $x = a^t$, we deduce that $t = \log_a x$ where the log is to base a, but this involves a much later way of thinking. Here we are really thinking of log as a function, while early workers in logarithms thought purely of the log as a number which aided calculation. It may have been Jacob Bernoulli who first understood the way that the log function is the inverse of the exponential function. On the other hand the first person to make the connection between logarithms and exponents may well have been James Gregory. In 1684 he certainly recognized the connection between logarithms and exponents, but he may not have been the first.

As far as we know the first time the number e appears in its own right is in 1690. In that year Leibniz wrote a letter to Huygens and in this he used the notation b for what we now call e. At last the number e had a name (even if not its present one) and it was recognized. Now the reader might ask, not unreasonably, why we have not started our article on the history of e at the point where it makes its first appearance. The reason is that although the work we have described previously never quite managed to identify e, once the number was identified then it was slowly realised that this earlier work is relevant. Retrospectively, the early developments on the logarithm became part of an understanding of the number e.

We mentioned above the problems arising from the fact that log was not thought of as a function. It would be fair to say that Johann Bernoull began the study of the calculus of the exponential function in 1697 when he published *Principia calculi exponentialum seu percurrentiuw*. The work involves the calculation of various exponential series and many results are achieved with term by term integration.

So much of our mathematical notation is due to Euler that it will come as no surprise to find that the notation e for this number is due to him. The claim which has sometimes been made, however, that Euler used the letter e because it was the first letter of his name is ridiculous. It is probably not even the case that the e comes from "exponential", but it may have just be the next vowel after "a" and Euler was already using the notation "a" in his work. Whatever the reason, the notation e made its first appearance in a letter Euler wrote to Goldbach in 1731. He made various discoveries regarding e in the following years, but it was not until 1748 when Euler published *Introduction in Analysin infinitorum* that he gave a full treatment of the ideas surrounding e. He showed that

$$e = 1 + \frac{1}{1!} + \frac{1}{2!} + \frac{1}{3!} + \cdots$$

and that e is the limit of $\left(1 + \frac{1}{n}\right)^n$ as n tends to infinity. Euler gave an approximation for e to 18 decimal places

$$e = 2.718\ 281\ 828\ 459\ 045\ 235$$

without saying where this came from. It is likely that he calculated the value himself, but if so there is no indication of how this was done. In fact taking about 20 terms of $1 + \frac{1}{1!} + \frac{1}{2!} + \frac{1}{3!} + \cdots$ will give the accuracy which Euler gave. Among other interesting results in this work is the connection between the sine and cosine functions and the complex exponential function, which Euler deduced using De Moivre's formula.

Interestingly Euler also gave the continued fraction expansion of e

and noted a pattern in the expansion. In particular he gave

$$\frac{e-1}{2} = \cfrac{1}{1 + \cfrac{1}{6 + \cfrac{1}{10 + \cfrac{1}{14 + \cfrac{1}{18 + \cdots}}}}}$$

and

$$e - 1 = 1 + \cfrac{1}{1 + \cfrac{1}{2 + \cfrac{1}{1 + \cfrac{1}{1 + \cfrac{1}{4 + \cfrac{1}{1 + \cfrac{1}{1 + \cfrac{1}{6 + \cdots}}}}}}}}$$

Euler did not give a proof that the patterns he spotted continue (which they do) but he knew that if such a proof were given it would prove that e is irrational. For, if the continued fraction for $\frac{e-1}{2}$ were to follow the pattern shown in the first few terms, 6, 10, 14, 18, 22, 26, ... (add 4 each time) then it will never terminate so $\frac{e-1}{2}$ (and so e) cannot be rational. One could certainly see this as the first attempt to prove that e is not rational.

The same passion that drove people to calculate to more and more decimal places of π never seemed to take hold in quite the same way for e. There were those who did calculate its decimal expansion, however, and the first to give e to a large number of decimal places was Shanks in 1854. It is worth noting that Shanks was an even more enthusiastic calculator of the decimal expansion of π. Glaisher showed that the first 137 places of Shanks calculations for e were correct but found an error which, after correction by Shanks, gave e to 205 places. In fact one needs about 120 terms of $1 + \frac{1}{1!} + \frac{1}{2!} + \frac{1}{3!} + \cdots$ to obtain e correct to 200 places.

Further calculations of decimal expansions followed. In 1884 Doorman calculated e to 346 places and found that his calculation agreed with that of Shanks as far as place 187 but then became different. In 1887 Adams calculated the base 10 log of e to 272 places.

Words and Expressions

prompt $adj.$ 迅速的,敏捷的;即时的,立刻的 $vt.$ 激励,鼓动 $n.$ 提示
appendix $n.$ 附录,附属物
occurrence $n.$ 发生;事件
explicitly $adv.$ 明确地
rectangular $adj.$ 矩形的,长方形的,直角的
rectangular hyperbola 等轴双曲线,直角双曲线
terminology $n.$ 术语,专门名词
dubious $adj.$ 怀疑的,可疑的
elusively $adv.$ 难懂地,令人困惑地
process $n.$ 过程;方法,步骤
limiting process 极限过程
deduce $vt.$ 演绎,推演
log funtion 对数函数
exponential function 指数函数
exponential series 指数级数
retrospectively $adv.$ 回顾
accuracy $n.$ 精度,精确度;准确,准确性
sine $n.$ 正弦
sine function 正弦函数
cosine $n.$ 余弦
cosine function 余弦函数

complex exponential function 复指数函数
De Moivre's formula 棣莫弗公式
terminate *adj*. 有限的,有结尾的 *vt*. 终止,结束 *vi*. 结束,结局
enthusiastic *adj*. 热心的,满腔热忱的
enthusiastic calculator 满腔热忱的计算者

1.3 INEQUALITIES FOR CONVEX FUNCTION

Convex functions

Convex functions are powerful tools for proving a large class of inequalities. They provide an elegant and unified treatment of the most important classical inequalities.

A real-valued function on an interval I is called *convex* if
$$f(\lambda x + (1-\lambda)y) \leq \lambda f(x) + (1-\lambda)f(y) \tag{1}$$
for every $x, y \in I$ and $\lambda \in [0,1]$; it is called *strictly convex* if
$$f(\lambda x + (1-\lambda)y) < \lambda f(x) + (1-\lambda)f(y) \tag{2}$$
for every $x, y \in I, x \neq y$ and $\lambda \in (0,1)$.

Notice

f is called *concave* (*strictly concave*) on I if $-f$ is *convex* (*strictly convex*) on I.

The geometrical meaning of convexity is clear: f is strictly convex if and only if for every two points $P = (x, f(x))$ and $Q = (y, f(y))$ on the graph of f, the point $R = (z, f(z))$ lies below the segment PQ for every z between x and y.

How to recognize a convex function without the graph? We can use (1) directly, but the following criterion is often very useful:

Test for Convexity

Let f be a twice differentiable function on I. Then f is convex on I

if $f''(x) \geq 0$ for every $x \in I$. f is strictly convex on I if $f''(x) > 0$ for every x in the interior of I.

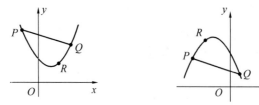

Figure 1 **Figure 2**

Remark

If f is a continuous function on I, then it can be proved that f is convex if and only if for all $x_1, x_2 \in I$

$$f\left(\frac{x_1 + x_2}{2}\right) \leq \frac{f(x_1) + f(x_2)}{2}$$

and it is strictly convex if and only if

$$f\left(\frac{x_1 + x_2}{2}\right) < \frac{f(x_1) + f(x_2)}{2}$$

for all $x_1, x_2 \in I$, $x_1 \neq x_2$.

Here are some basic examples of strictly convex functions:

(i) $f(x) = x^{2n}$, $x \in \mathbf{R}$ and n is a positive integer;

(ii) $f(x) = x^p$, $x \geq 0, p > 1$;

(iii) $f(x) = \dfrac{1}{(x+a)^p}$, $x > -a, p > 0$;

(iv) $f(x) = \tan x$, $x \in [0, \dfrac{\pi}{2}]$;

(v) $f(x) = e^x$, $x \in \mathbf{R}$.

The following are examples of strictly concave functions:

(i) $f(x) = \sin x$, $x \in [0, \pi]$;

(ii) $f(x) = \cos x$, $x \in [-\dfrac{\pi}{2}, \dfrac{\pi}{2}]$;

(iii) $f(x) = \ln x, x \in (0, \infty)$;
(iv) $f(x) = x^p, x \geq 0, p \in (0, 1)$.

Notice

1. The linear function $f(x) = ax + b, x \in \mathbf{R}$ is convex and also concave.

2. The sum of two convex(concave) functions is a convex(concave) function.

Jensen's Inequality

Jensen's inequality is an extension of (1). It was named after the Danish mathematician who proved it in 1905.

Jensen's inequality

Let $f: I \to \mathbf{R}$ be a convex function. Let $x_1, \cdots, x_n \in I$ and $\lambda_1, \cdots, \lambda_n \geq 0$ such that $\lambda_1 + \lambda_2 + \cdots + \lambda_n = 1$. Then

$$f(\lambda_1 x_1 + \lambda_2 x_2 + \cdots + \lambda_n x_n) \leq \lambda_1 f(x_1) + \cdots + \lambda_n f(x_n) \quad (3)$$

Proof Let's use mathematical induction. The inequality is true for $n = 1$. Now assume that it is true for $n = k$, and let's show that it remains true for $n = k + 1$.

Let $x_1, \cdots, x_k, x_{k+1} \in I$ and let $\lambda_1, \cdots, \lambda_k, \lambda_{k+1} \geq 0$ with $\lambda_1 + \lambda_2 + \cdots + \lambda_k + \lambda_{k+1} = 1$. At least one of $\lambda_1, \lambda_2, \cdots, \lambda_{k+1}$ must be less than 1 (otherwise the inequality is trivial). Without loss of generality, let

$$\lambda_{k+1} < 1 \text{ and } u = \frac{\lambda_1}{1 - \lambda_{k+1}} x_1 + \cdots + \frac{\lambda_k}{1 - \lambda_{k+1}} x_k$$

We have

$$\frac{\lambda_1}{1 - \lambda_{k+1}} + \cdots + \frac{\lambda_k}{1 - \lambda_{k+1}} = 1$$

and also

$$\lambda_1 x_1 + \cdots + \lambda_k x_k + \lambda_{k+1} x_{k+1} = (1 - \lambda_{k+1}) u + \lambda_{k+1} x_{k+1}$$

Now, since f is convex

$$f((1-\lambda_{k+1})u + \lambda_{k+1}x_{k+1}) \leq (1-\lambda_{k+1})f(u) + \lambda_{k+1}f(x_{k+1})$$

and, by our induction hypothesis

$$f(u) \leq \frac{\lambda_1}{1-\lambda_{k+1}}f(x_1) + \cdots + \frac{\lambda_k}{1-\lambda_{k+1}}f(x_k)$$

Hence, combining the above two inequalities, we get

$$f(\lambda_1 x_1 + \cdots + \lambda_{k+1}x_{k+1}) \leq \lambda_1 f(x_1) + \cdots + \lambda_{k+1}f(x_{k+1})$$

Thus, the inequality is established for $n = k + 1$, and therefore, by mathematical induction, it holds for any positive integer n.

Remarks

1. For strictly convex functions, the equality in (3) holds if and only if $x_1 = x_2 = \cdots = x_n$. Use mathematical induction to prove it.

2. If $\lambda_1 = \lambda_2 = \cdots = \lambda_n = \frac{1}{n}$, then (3) becomes

$$f\left(\frac{x_1 + x_2 + \cdots + x_n}{n}\right) \leq \frac{f(x_1) + f(x_2) + \cdots + f(x_n)}{n} \quad (4)$$

3. If f is a concave function, then (3) and (4) read as

$$f(\lambda_1 x_1 + \cdots + \lambda_n x_n) \geq \lambda_1 f(x_1) + \cdots + \lambda_n f(x_n) \quad (3')$$

and

$$f\left(\frac{x_1 + x_2 + \cdots + x_n}{n}\right) \geq \frac{f(x_1) + f(x_2) + \cdots + f(x_n)}{n} \quad (4')$$

Jensen's Inequality has variety of applications. It can be used to prove many of the most important classical inequalities.

Weighted AM – GM Inequality

Let $x_1, x_2, \cdots, x_n \geq 0$, $\lambda_1, \cdots, \lambda_n > 0$ such that $\lambda_1 + \cdots + \lambda_n = 1$. Then

$$\lambda_1 x_1 + \cdots + \lambda_n x_n \geq x_1^{\lambda_1} x_2^{\lambda_2} \cdots x_n^{\lambda_n} \quad (5)$$

The equality holds if and only if $x_1 = x_2 = \cdots = x_n$.

Proof We may assume that $x_1, x_2, \cdots, x_n > 0$. Let $f(x) = \ln x$,

$x \in (0, \infty)$. Since f is strictly concave on $(0, \infty)$, by using $(3')$ we get
$$\ln(\lambda_1 x_1 + \cdots + \lambda_n x_n) \geq \lambda_1 \ln x_1 + \cdots + \lambda_n \ln x_n$$
or, equivalently, $\ln(\lambda_1 x_1 + \cdots + \lambda_n x_n) \geq \ln(x_1^{\lambda_1} \cdots x_n^{\lambda_n})$, and hence
$$\lambda_1 x_1 + \cdots + \lambda_n x_n \geq x_1^{\lambda_1} x_2^{\lambda_2} \cdots x_n^{\lambda_n}$$
(since $f(x) = \ln x$ is a strictly increasing function).

By taking $\lambda_1 = \lambda_2 = \cdots = \lambda_n = \dfrac{1}{n}$ in (5), we obtain AM – GM Inequality.

AM – GM Inequality

If $x_1, x_2, \cdots, x_n \geq 0$, then
$$\frac{x_1 + x_2 + \cdots + x_n}{n} \geq \sqrt[n]{x_1 x_2 \cdots x_n} \tag{6}$$
with equality if and only if $x_1 = x_2 = \cdots = x_n$.

Let $x_1, x_2, \cdots, x_n, \lambda_1, \lambda_2, \cdots, \lambda_n > 0$ be such that $\lambda_1 + \cdots + \lambda_n = 1$. For each $t \in \mathbf{R}, t \neq 0$, the weighted mean M_t of order t is defined as
$$M_t = (\lambda_1 x_1^t + \cdots + \lambda_n x_n^t)^{\frac{1}{t}}$$
Some particular situations are significant
$$M_1 = \lambda_1 x_1 + \cdots + \lambda_n x_n$$
is called the *weighted arithmetic mean* (WAM); and
$$M_{-1} = \frac{1}{\dfrac{\lambda_1}{x_1} + \dfrac{\lambda_2}{x_2} + \cdots + \dfrac{\lambda_n}{x_n}}$$
is called the *weighted harmonic mean* (WHM)
$$M_2 = \sqrt{\lambda_1 x_1^2 + \cdots + \lambda_n x_n^2}$$
is called the *weighted root mean square* (WRMS).

It can be shown by using L'Hopital's Rule that
$$\lim_{t \to 0} M_t = x_1^{\lambda_1} x_2^{\lambda_2} \cdots x_n^{\lambda_n}$$
So, if we denote $M_0 = \lim\limits_{t \to 0} M_t$, we see that

$$M_0 = x_1^{\lambda_1} x_2^{\lambda_2} \cdots x_n^{\lambda_n}$$

which is called the *weighted geometric mean* (WGM).

Also, if we set

$$M_\infty = \lim_{t \to \infty} M_t \text{ and } M_{-\infty} = \lim_{t \to -\infty} M_t$$

we obtain

$$M_\infty = \max\{x_1, \cdots, x_n\}, M_{-\infty} = \min\{x_1, \cdots, x_n\}$$

Power Mean Inequality

Let $x_1, x_2, \cdots, x_n, \lambda_1, \lambda_2, \cdots, \lambda_n > 0$ be such that $\lambda_1 + \cdots + \lambda_n = 1$. If t and s are non-zero real numbers such that $s < t$, then

$$(\lambda_1 x_1^s + \cdots + \lambda_n x_n^s)^{\frac{1}{s}} \leq (\lambda_1 x_1^t + \cdots + \lambda_n x_n^t)^{\frac{1}{t}} \quad (7)$$

Proof If $0 < s < t$ or $s < 0 < t$, the inequality (7) is obtained by applying Jensen's Inequality (3) to the strictly convex function $f(x) = x^{\frac{t}{s}}$. Indeed, if $a_1, a_2, \cdots, a_n, \lambda_1, \cdots, \lambda_n > 0$ and $\lambda_1 + \cdots + \lambda_n = 1$, then

$$(\lambda_1 a_1 + \lambda_2 a_2 + \cdots + \lambda_n a_n)^{\frac{t}{s}} \leq \lambda_1 a_1^{t/s} + \lambda_2 a_2^{t/s} + \cdots + \lambda_n a_n^{t/s}$$

By choosing $a_1 = x_1^s, \cdots, a_n = x_n^s$, we immediately obtain (7).

If $s < t < 0$, then $0 < -t < -s$, and by applying (7) for $\frac{1}{x_1}, \frac{1}{x_2}, \cdots, \frac{1}{x_n}$, we get

$$\left(\lambda_1 \left(\frac{1}{x_1}\right)^{-t} + \cdots + \lambda_n \left(\frac{1}{x_n}\right)^{-t}\right)^{\frac{1}{-t}} \leq \left(\lambda_1 \left(\frac{1}{x_1}\right)^{-s} + \cdots + \lambda_n \left(\frac{1}{x_n}\right)^{-s}\right)^{\frac{1}{-s}}$$

which can be rewritten as (7).

Remarks

If $t < 0 < s$, then $M_t \leq M_0 \leq M_s$. Also, we have the following classical inequality

$$M_{-\infty} \leq M_{-1} \leq M_0 \leq M_1 \leq M_2 \leq M_\infty$$

Hölder's Inequality

If $p, q > 1$ are real numbers such that $\dfrac{1}{p} + \dfrac{1}{q} = 1$ and $a_1, \cdots, a_n$, $b_1, \cdots, b_n$ are real (complex) numbers, then

$$\sum_{k=1}^{n} |a_k| |b_k| \leq \left(\sum_{k=1}^{n} |a_k|^p \right)^{\frac{1}{p}} \left(\sum_{k=1}^{n} |b_k|^q \right)^{\frac{1}{q}} \quad (8)$$

Proof We may assume that $|a_k| > 0$, $K = 1, \cdots n$. The function $f(x) = x^q$ is strictly convex on $(0, \infty)$, hence, by Jensen's inequality

$$\left(\sum_{k=1}^{n} \lambda_k x_k \right)^q \leq \sum_{k=1}^{n} \lambda_k x_k^q$$

where $x_1, \cdots, x_n, \lambda_1, \cdots, \lambda_n > 0$ and $\lambda_1 + \cdots + \lambda_n = 1$. Let $A = \sum_{k=1}^{n} |a_k|^p$. By choosing $\lambda_k = \dfrac{1}{A} |a_k|^p$ and $x_k = \dfrac{1}{\lambda_k} |a_k| |b_k|$ in the above inequality, we obtain (8).

Remarks

1. In (8), the equality holds if and only if $x_1 = x_2 = \cdots = x_n$. That is

$$\frac{|a_1|^p}{|b_1|^q} = \frac{|a_2|^p}{|b_2|^q} = \cdots = \frac{|a_n|^p}{|b_n|^q}$$

Notice that this chain of equalities is taught in the following way: if a certain $b_k = 0$, then we should have $a_k = 0$.

2. If $p = q = 2$, Hölder's Inequality is just **Cauchy's Inequality**

$$\left(\sum_{k=1}^{n} |a_k| |b_k| \right)^2 \leq \left(\sum_{k=1}^{n} |a_k|^2 \right) \left(\sum_{k=1}^{n} |b_k|^2 \right)$$

The equality occurs when

$$\frac{|a_1|}{|b_1|} = \frac{|a_2|}{|b_2|} = \cdots = \frac{|a_n|}{|b_n|}$$

Minkowski's Triangle Inequality

If $p > 1$ and $a_1, a_2, \cdots, a_n, b_1, b_2, \cdots, b_n \geq 0$, then

$$\left(\sum_{k=1}^{n}(a_k + b_k)^p\right)^{\frac{1}{p}} \leq \left(\sum_{k=1}^{n} a_k^p\right)^{\frac{1}{p}} + \left(\sum_{k=1}^{n} b_k^p\right)^{\frac{1}{p}} \qquad (9)$$

Proof We may assume $a_k > 0$, $k = 1, \cdots, n$. The function $f(x) = (1 + x^{\frac{1}{p}})^p$, $x \in (0, \infty)$, is strictly concave since $f''(x) = \dfrac{1-p}{p}(1 + x^{\frac{1}{p}})^{p-2} \cdot x^{\frac{1}{p}-2} < 0$. By Jensen's Inequality

$$\left[1 + \left(\sum_{k=1}^{n} \lambda_k x_k\right)^{\frac{1}{p}}\right]^p \geq \sum_{k=1}^{n} \lambda_k (1 + x_k^{1/p})^p$$

where $x_1, \cdots, x_n, \lambda_1, \cdots, \lambda_n > 0$ and $\lambda_1 + \cdots + \lambda_n = 1$. Let $A = \sum_{k=1}^{n} a_k^p$. By taking $\lambda_k = \dfrac{a_k^p}{A}$ and $x_k = \dfrac{b_k^p}{a_k^p}$ for $k = 1, \cdots, n$ in the above inequality, we obtain (9).

Remarks

1. The equality in (9) occurs if and only if

$$\frac{b_1}{a_1} = \frac{b_2}{a_2} = \cdots = \frac{b_n}{a_n}$$

2. If $p = 2$ we get the so-called Triangle Inequality:

$$\sqrt{\sum_{k=1}^{n}(a_k + b_k)^2} \leq \sqrt{\sum_{k=1}^{n} a_k^2} + \sqrt{\sum_{k=1}^{n} b_k^2}$$

Example 1 If $a, b \geq 0$ and $a + b = 2$, then

$$(1 + \sqrt[5]{a})^5 + (1 + \sqrt[5]{b})^5 \leq 2^6$$

Solution Since $f(x) = (1 + \sqrt[5]{x})^5$ is strictly concave on $[0, \infty)$, by using Jensen's Inequality (4') we get

$$2\left(1 + \sqrt[5]{\frac{a+b}{2}}\right)^5 \geq (1 + \sqrt[5]{a})^5 + (1 + \sqrt[5]{b})^5$$

By substituting $a + b = 2$, we get the required inequality. The equality occurs when $a = b = 1$.

Example 2 If $a, b, c > 0$, then
$$a^a \cdot b^b \cdot c^c \geq \left(\frac{a+b+c}{3}\right)^{a+b+c}$$

Solution The above inequality is equivalent to
$$\ln(a^a \cdot b^b \cdot c^c) \geq \ln\left(\frac{a+b+c}{3}\right)^{a+b+c}$$
$$a\ln a + b\ln b + c\ln c \geq (a+b+c)\ln\left(\frac{a+b+c}{3}\right)$$

Let $f(x) = x\ln x$, $x \in (0, \infty)$. Since $f''(x) = \frac{1}{x} > 0$, the function f is strictly convex on $(0, \infty)$. Now, the above inequality follows from (4).

Example 3 If $a, b, c > 0$ then
$$\frac{a}{a+3b+3c} + \frac{b}{3a+b+3c} + \frac{c}{3a+3b+c} \geq \frac{3}{7}$$

Solution Let s be a positive number and $f(x) = \frac{x}{s-x} = \frac{s}{s-x} - 1$, $x \in (0, s)$. The function f is strictly convex since $f''(x) = \frac{2s}{(s-x)^3} > 0$. We get
$$\frac{2a}{s-2a} + \frac{2b}{s-2b} + \frac{2c}{s-2c} \geq 3 \frac{\frac{1}{3}(2a+2b+2c)}{s - \frac{1}{3}(2a+2b+2c)}$$

or
$$\frac{a}{s-2a} + \frac{b}{s-2b} + \frac{c}{s-2c} \geq \frac{3(a+b+c)}{3s - 2(a+b+c)}$$

If we take $s = 3(a+b+c)$, the required inequality follows.

Example 4 If $a_1, a_2, \cdots, a_n \geq 1$, then
$$\sum_{k=1}^{n} \frac{1}{1+a_k} \geq \frac{n}{1 + \sqrt[n]{a_1, a_2, \cdots, a_n}}$$

Solution Let $f(x) = \frac{1}{1+e^x}$, $x \in [0, \infty)$. The function f is strictly convex since $f''(x) = \dfrac{e^x(e^x - 1)}{(e^x + 1)^3} > 0$ on $(0, \infty)$. Using (4), we get

$$\sum_{k=1}^{n} \frac{1}{1 + e^{x_k}} \geq \frac{n}{1 + e^{\frac{1}{n}\sum_{k=1}^{n} x_k}}$$

By taking $x_k = \ln a_k$, $k = 1, \cdots, n$, we obtain the required inequality.

Example 5 For a triangle with angles α, β and γ, the following inequalities hold:

- $\sin\alpha + \sin\beta + \sin\gamma \leq \dfrac{3\sqrt{3}}{2}$;

- $\sqrt{\sin\alpha} + \sqrt{\sin\beta} + \sqrt{\sin\gamma} \leq 3\sqrt[4]{\dfrac{3}{4}}$;

- $\sin\alpha \cdot \sin\beta \cdot \sin\gamma \leq \dfrac{3\sqrt{3}}{8}$;

- $\cos\alpha \cdot \cos\beta \cdot \cos\gamma \leq \dfrac{1}{8}$;

- $\sec\dfrac{\alpha}{2} + \sec\dfrac{\beta}{2} + \sec\dfrac{\gamma}{2} \geq 2\sqrt{3}$.

Solution Use the Jensen Inequality for the strictly concave functions $\sin x$, $\sqrt{\sin x}$, $\ln \sin x$, $\ln \cos x$, and for the strictly convex function $\sec \dfrac{x}{2}$, $x \in (0, \pi)$.

Example 6 Let $a_1, \cdots, a_n, \lambda_1, \cdots, \lambda_n > 0$ and $\lambda_1 + \cdots + \lambda_n = 1$. If $a_1^{\lambda_1} \cdots a_n^{\lambda_n} = 1$, then

$$a_1 + a_2 + \cdots + a_n \geq \frac{1}{\lambda_1^{\lambda_1} \cdots \lambda_n^{\lambda_n}}$$

The equality occurs if and only if $a_k = \dfrac{\lambda_k}{\lambda_1^{\lambda_1} \cdots \lambda_n^{\lambda_n}}$ for $k = 1, \cdots, n$.

Solution By using the Weighted AM − GM Inequality, we have

$$a_1 + \cdots + a_n = \lambda_1 \left(\frac{a_1}{\lambda_1}\right) + \cdots + \lambda_n \left(\frac{a_n}{\lambda_n}\right) \geq$$

$$\left(\frac{a_1}{\lambda_1}\right)^{\lambda_1} \cdots \left(\frac{a_n}{\lambda_n}\right)^{\lambda_n} = \frac{1}{\lambda_1^{\lambda_1} \cdots \lambda_n^{\lambda_n}}$$

The equality occurs if and only if
$$\frac{a_1}{\lambda_1} = \frac{a_2}{\lambda_2} = \cdots = \frac{a_n}{\lambda_n}$$
in which case the constraint $a_1^{\lambda_1} \cdots a_n^{\lambda_n} = 1$ leads to $a_k = \frac{\lambda_k}{\lambda_1^{\lambda_1} \cdots \lambda_n^{\lambda_n}}$ for $k = 1, \cdots, n$.

Example 7 If $a_1, \cdots, a_n > 0$ and $a_1 a_2 \cdots a_n = 1$, then
$$a_1 + 2\sqrt{a_2} + \cdots + n\sqrt[n]{a_n} \geq \frac{n(n+1)}{2}$$

Hint: Use the Weighted AM-GM Inequality.

Example 8

(i) If $a, b, c > 0$, then
$$\frac{a^{10} + b^{10} + c^{10}}{a^5 + b^5 + c^5} \geq \left(\frac{a+b+c}{3}\right)^5$$

(ii) If $a_1, a_2, \cdots, a_n > 0$ and $k > p \geq 0$, then
$$\frac{a_1^k + \cdots + a_n^k}{a_1^p + \cdots + a_n^p} \geq \left(\frac{a_1 + \cdots + a_n}{n}\right)^{k-p}$$

Solution (i) A particular case of (ii).

(ii) If $M_t = \left(\frac{a_1^t + \cdots + a_n^t}{n}\right)^{\frac{1}{t}}$. Then, by using the Power Mean Inequality (7), we get
$$a_1^k + a_2^k + \cdots + a_n^k = n M_k^k = n M_k^p M_k^{k-p} \geq$$
$$n M_p^p M_1^{k-p} \text{ by}(7) =$$
$$(a_1^p + \cdots + a_n^p) \cdot \left(\frac{a_1 + \cdots + a_n}{n}\right)^{k-p}$$

Words and Expressions

inequality *n*. 不等式,不等
convex *n*. 凸 *adj*. 凸的
function *n*. 函数;函词,函项
convex function 凸函数

real *adj.* 实的,实数的,有效的 *n.* 实数,实部,实型
value *n.* 值,数值
real-valued function 实值函数
interval *n.* 区间
strictly *adv.* 严格地
strictly convex 严格凸函数
geometrical *adj.* = geometric 几何的,几何学的
meaning *n.* 意义,含义
geometrical meaning 几何含义
graph *n.* 图,图形;网络
segment *n.* 段,节
criterion *n. pl.* = *criteria* 准则,判别准则,判据
convexity *n.* 凸性
differentiable *adj.* 可微的
differentiable function 微分方程
twice differentiable function 二阶微分方程
continuous *adj.* 连续的
continuous function 连续函数
positive *adj.* 正的,肯定的
integer *n.* 整数
positive integer 正整数
linear *adj.* 线性的,一次的
linear function 线性函数,一次函数
generality *adj.* 普遍性,普通性,一般性
without loss of generality 不失一般性
induction *n.* 归纳,归纳法
hypothesis *n.* 假设
induction hypothesis 归纳假设
mathematical *adj.* 数学的
mathematical induction 数学归纳法

weighted *adj.* 加权的
AM arithmetic mean 算术平均值
GM geometric mean 几何平均值
weighted AM – GM Inequality 加权算术平均几何平均不等式
if and only if 当且仅当
equivalently *adv.* 等价地,等势地,相等地
strictly increasing function 严格增函数
weighted mean 加权平均
weighted arithmetic mean（WAM） 加权算术平均
harmonic *n.* 调和,调和函数 *adj.* 调和的
harmonic mean 调和平均,调和中项,调和中数
weighted harmonic mean（WHM） 加权调和平均
root *n.* 根
root mean square 均方根
weighted root mean square（WRMS） 加权均方根
geometric mean 几何平均
weighted geometric mean（WGM） 加权几何平均
power *n.* 幂,乘方;势;权
power mean 幂平均
power mean inequality 幂平均不等式
non-zero 非零
non-zero real number 非零实数
such that 使得
angle *n.* 角
triangle *n.* 三角,三角形
triangle inequality 三角不等式

2

The History

2.1 THE INTEGRAL

The path to the development of the integral is a branching one, where similar discoveries were made simultaneously by different people. The history of the technique that is currently known as integration began with attempts to find the area underneath curves. The foundations for the discovery of the integral were first laid by Cavalieri, an Italian Mathematician, in around 1635. Cavalieri's work centered around the observation that a curve can be considered to be sketched by a moving point and an area to be sketched by a moving line.

Cavalieri's Method of Indivisbles

In order to deal with the geometrical notion of a moving point, Cavalieri worked with what he called "indivisibles". That is, if a moving point can be considered to sketch a curve, then Cavalieri viewed the curve as the sum of its points. By this notion, each curve is made up of an infinite number of points, or "indivisibles". Likewise, the "indivisibles" that composed an area were an infinite number of lines. Though Cavalieri was not the first person to consider geometric figures in terms of the infinitesimal (Kepler had done so before him), he was the first to use such a notion in the computation of areas.

In order to introduce Cavalieri's method, consider finding the area of a triangle. For many years, it had been known that the area of a triangle was $\frac{1}{2}$ the area of a rectangle which has the same base and height.

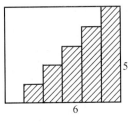

Figure 3

In Figure 3, the rectangle has a base of 6 units and a height of 5 units ($A = bh$, so the total area is 30 units). The total area of the inner rectangular regions can easily be computed by taking the sum of all the inner rectangles. Comparing the two areas

$$\frac{\text{Area of shaded region}}{\text{Area of entire rectangle}} = \frac{0+1+2+3+4+5}{5 \times 6} = \frac{15}{30} = \frac{1}{2}$$

Using the same methodology, the ratio for a larger rectangle with a greater number of inner subdivisions is computed

$$\frac{\text{Area of shaded region}}{\text{Area of entire rectangle}} = \frac{0+1+2+3+\cdots+10}{10 \times 11} = \frac{55}{110} = \frac{1}{2}.$$

The total area of the inner regions is always one-half the area of the total rectangle. This can be shown formally by using the closed form of the summation for the numerator

$$\sum_{i=1}^{n} i = 0+1+2+\cdots+n = \frac{n(n+1)}{2}$$

Using the closed form, it can be seen that

$$\frac{\text{Area of shaded region}}{\text{Area of entire rectangle}} = \frac{\sum_{i=1}^{n} i}{n(n+1)} = \frac{\frac{1}{2}n(n+1)}{n(n+1)} = \frac{1}{2}$$

Cavalieri now took a step of great importance to the formation of the integral calculus. He utilized his notion of "indivisibles" to imagine that there were an infinite number of shaded regions. He saw that as the individual shaded regions became small enough to simply be lines, the jagged steps would gradually define a line. As the jagged steps became a line, the shaded region would form a triangle. As the number of shaded regions increases, the ratio remains simply one-half.

Cavalieri's methodology agreed with the long-held result that the area of a triangle was one-half the product of the base and height. He had also shown that his notion of "indivisibles" can be used to successfully describe the area underneath the curve. That is, as the areas of the rectangles turn into lines, their sum does indeed produce the area underneath the curve (in this case, a line). Cavalieri went on to use his method of "indivisibles" to find the area underneath many different curves. However, he was never able to formulate his techniques into a logically consistent foundation that others accepted. Though Cavalieri's techniques clearly worked, it was not until Sir John Wallis of England that the limit was formally introduced in 1656 and the foundation for the integral calculus was solidified.

In order to fully understand Wallis' contributions to the integral calculus, it is first necessary to see how Cavalieri's theoretical techniques can be applied to find the area underneath a curve more complicated than a line. In order to do so, this technique will be applied to find the area underneath the parabola $y = x^2$.

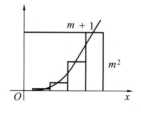

Figure 4

Each rectangular region has a base of 1 unit along the x-axis and height of x^2 (obtained from the definition of the parabola). The number of rectangular regions will be defined to be m. Cavalieri again attempted to express the area underneath the curve as the ratio of an area that was already known. He considered the area enclosing all of the m rectangles. It can easily be seen from the diagram that the base of this rectangle will be $m + 1$ (there are m rectangles, the first starting at $\frac{1}{2}$ and the last one ending at $m + \frac{1}{2}$. The height of the enclosing rectangle will be m^2, from the definition of the parabola. The ratio can now be expressed with the following equation

$$\frac{\text{Total area of } m \text{ rectangles}}{\text{Area of bounding rectangle}} = \frac{1^2 + 2^2 + 3^2 + \cdots + m^2}{(m+1)m^2}$$

Recall that the area of a rectangle is defined by the product of its base and height. It was stated that the bounding rectangle had a base of $m + 1$ and a height of m^2, which accounts for the denominator. The numerator is easily explained as well: each of the m rectangles has a base of 1 and a height of its x value squared. Cavalieri now proceeded to calculate the ratio for different values of m. In doing so, he noticed a pattern and was able to establish a closed form for the ratio of the areas

$$\frac{\text{Total area of } m \text{ rectangles}}{\text{Area of bounding rectangle}} = \frac{1}{3} + \frac{1}{6m}$$

Cavalieri then utilized his important principle of "indivisibles" to make another important leap in the development of the calculus. He noticed that as he let m grow larger, the term had less influence on the outcome of the result. In modern terms, he noticed that

$$\lim_{m \to \infty} \left(\frac{1}{3} + \frac{1}{6m} \right) = \frac{1}{3}$$

That is, as he lets the number of rectangles grow to infinity, the ratio of the areas will become closer to $\frac{1}{3}$. Though Cavalieri did not formally introduce the notation for limits, he did utilize the idea in the computation of areas. After using the concept of infinity to describe the ratios of the area, he was able to derive an algebraic expression for the area underneath the parabola. For at any distance x along the x-axis, the height of the parabola would be x^2. Therefore, the area of the rectangle enclosing the rectangular subdivisions at a point x was equal to $x(x^2)$ or x^3. From his earlier result, the area underneath the parabola is equal to $\frac{1}{3}$ the area of the bounding rectangle. In other words

$$Area\ under\ x^2 = \frac{1}{3}x^3$$

With this technique, Cavalieri had laid the fundamental building block for integration.

Wallis' Law for Integration of Polynomials

John Wallis' contribution to the integral calculus was to derive an algebraic law for integration that alleviated the necessity of going through such analysis for each curve. Through examining the relationship between a function and the function that describes its area (henceforth referred to as the area-function), he was able to derive an algebraic law for determining area-functions. Rather than simply present the algebraic relationship (which the reader is doubtless familiar with if (s)he has studied a minimal amount of calculus), we will perform a similar analysis as to what led Wallis to derive his law.

First, consider the graph of the function $y = k$ or $y = kx^0$ (Firgue 5):

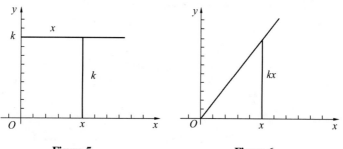

Figure 5   Figure 6

Clearly, it can be seen from the diagram, that the area underneath the line at any point along the x-axis will be kx or $A = \frac{1}{1} kx$.

Next, consider the graph of the function $y = kx$ (Firgue 6):

At any point x along the x-axis, the height will be equal to kx. Since the area forms a triangle, the area underneath the curve can be expressed as $\frac{1}{2}$ the base times the height or $A = \frac{1}{2} kx^2$. As it was already shown above, the area underneath a parabola $y = kx^2$, can be expressed as $A = \frac{1}{3} kx^3$. Wallis noticed an algebraic relationship between a function and its associated area-function. That is, the area-function of $y = kx^n$ is $A = \frac{1}{n+1} kx^{n+1}$. Wallis went on to show that not only does this hold true where n is a natural number (which had been the extent of Cavalieri's work), but that it also worked for negative and fractional exponents. Wallis also showed that the area underneath a polynomial composed of terms with different exponents (e.g. $y = 4x^3 + 3x^2 + x + 1$) can be computed by using his law on each of the terms independently.

Fermat's Approach to Integration

One of the first major uses of infinite series in the development of calculus came from Pierre De Fermat's method of integration. Though previous methods of integration had used the notion of infinite lines describing an area, Fermat was the first to use infinite series in his methodology. The first step in his method involved a unique way of describing the infinite rectangles making up the area under a curve.

Fermat noticed that by dividing the area underneath a curve into successively smaller rectangles as x became closer to zero, an infinite number of such rectangles would describe the area precisely. His methodology was to choose a value $0 < e < 1$, such that a rectangle was formed underneath the curve $y = x^{\frac{p}{q}}$ at each power of e times x (see Figure 7, **NOTE**: e was simply Fermat's choice of variable names, not $e = 2.71828\cdots$). Fermat then computed each area individually

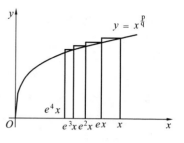

Figure 7

$$(x - ex)x^{\frac{p}{q}} = x(1-e)x^{\frac{p}{q}} = (1-e)x^{\frac{p+q}{q}}$$

$$(ex - e^2x)(ex)^{\frac{p}{q}} = ex(1-e)(ex)^{\frac{p}{q}} = (1-e)e^{\frac{p+q}{q}}x^{\frac{p+q}{q}}$$

$$(e^2x - e^3x)(e^2x)^{\frac{p}{q}} = e^2x(1-e)(e^2x)^{\frac{p}{q}} = (1-e)e^{2(\frac{p+q}{q})}x^{\frac{p+q}{q}}$$

The first equation represents the area of the largest rectangle, the second equation the next rectangle to the left, and so on. The areas are simply found by multiplying the base times the height. The base is known by the power of e, and the height by evaluating $y = x^{\frac{p}{q}}$ at the given x value. The simplifications of each area expression are given in a form that will be useful when attempting to find the infinite sum. Fermat's next step was to compute the infinite sum of these rectangles as the power of e approached infinity.

$$(1-e)x^{\frac{p+q}{q}} + (1-e)e^{\frac{p+q}{q}}x^{\frac{p+q}{q}} + (1-e)e^{2(\frac{p+q}{q})}x^{\frac{p+q}{q}} + \cdots$$

$$1\,term = (1-e)x^{\frac{p+q}{q}}(1)$$

$$2\,terms = (1-e)x^{\frac{p+q}{q}}(1 + e^{\frac{p+q}{q}})$$

$$3\,terms = (1-e)x^{\frac{p+q}{q}}(1 + e^{\frac{p+q}{q}} + e^{2(\frac{p+q}{q})})$$

...

$$n\,terms = (1-e)x^{\frac{p+q}{q}}(1 + e^{\frac{p+q}{q}} + e^{2(\frac{p+q}{q})} + e^{3(\frac{p+q}{q})} + \cdots)$$

By determining the sum of each increasing finite series, he was able to develop an expression for the infinite sum.

In order to find a closed form for the expression
$$(1 + e^{\frac{p+q}{q}} + e^{2(\frac{p+q}{q})} + e^{3(\frac{p+q}{q})} + \cdots)$$
note that the sum is a geometric series of the form
$$(1 + x + x^2 + x^3 + \cdots)$$
If $0 < x < 1$, the sum is $\frac{1}{1-x}$ (this can be shown to be true by long dividing $(1 - x)$ into 1). Therefore, by substituting $e^{\frac{p+q}{q}}$ back in for x and inserting into the overall equation, the area can be expressed as
$$A = \frac{1}{1 - e^{\frac{p+q}{q}}}(1 - e)x^{\frac{p+q}{q}}$$

Fermat now wished to express the area entirely in terms of x, and in order to do so substituted $e = E^q$, which by simplification and factoring out $(1 - E)$
$$A = \frac{(1 - E^q)x^{\frac{p+q}{q}}}{1 - E^{p+q}} = \frac{(1 - E)(1 + E + E^2 + \cdots + E^{q-1})x^{\frac{p+q}{q}}}{(1 - E)(1 + E + E^2 + \cdots + E^{p+q-1})} = \frac{(1 + E + E^2 + \cdots + E^{q-1})x^{\frac{p+q}{q}}}{(1 + E + E^2 + \cdots + E^{p+q-1})}$$

Fermat now made a step that with the benefit of current knowledge is explainable, but at that time was not properly justified. That is, Fermat said let $E = 1$ and since $E^q = 1^q = 1$ and because $E^q = e$ then e must also equal 1. By substituting 1 for E in the area expression above
$$A = \frac{(1 + 1 + 1^2 + \cdots + 1^{q-1})x^{\frac{p+q}{q}}}{(1 + 1 + 1^2 + \cdots + 1^{p+q-1})} = \left(\frac{q}{p+q}\right)x^{\frac{p+q}{q}}$$

Although this methodology yielded the appropriate result for the area underneath the curve, Fermat's justification of letting $E = 1$ was not properly formulated. What he actually was doing was taking the limit as E approaches 1 and as E approaches 1 so too will e. As e approaches 1, then e raised to any power will also approach 1, and the infinite sum of the areas underneath the curve has been determined. The notion of a limit was hinted at in Fermat's work, but it was not formally defined until later.

  Wallis and Fermat's work had laid the groundwork for the modern concept of the integral. However, what Fermat and Wallis had failed to recognize was the relationship between the differential and the integral. That idea would be developed simultaneously by two men: Newton and Leibniz. This would later be known as the Fundamental Theorem of Calculus and, as the name implies, it is a landmark discovery in the history of the Calculus. However, before proceeding on to describe this important theorem, it is first necessary to examine the development of the differential.

Words and Expressions

integral *n.* 积分,整数 *adj.* 积分的,正的
integration *n.* 积分,积分法
curve *n.* 曲线
method *n.* 法,方法
indivisible *adj.* 除不尽的
method of indivisibles 不可分量法
infinitesimal *n.* 无穷小,无限小 *adj.* 无穷小的,无限小的
rectangle *n.* 矩形(又称"长方形")
calculus *n.* 微积分,微积分学;演算
integral calculus 积分学
region *n.* 区域
shaded regions 阴影区域
product *n.* 积,乘积
base *n.* 底,基
height *n.* 高,高度
the product of the base and height 底与高之积
underneath *prep.* 在……之下

diagram $n.$ 图,图表,图解
parabola $n.$ 抛物线
denominator $n.$ 分母
numerator $n.$ 分子
infinity $n.$ 无穷大,无穷
algebraic $adj.$ 代数的
axis $n.$ 轴
x-axis x 轴
polynomial $n.$ 多项式
area $n.$ 面积
area-function 面积函数
relationship $n.$ 关系
algebraic relationship 代数关系
minimal $adj.$ 极小的
similar $adj.$ 相似的
analysis $n.$ 分析,分析学;解析
negative $adj.$ 否定的;负的
exponents $n.$ 指数,幂
approach $n.$ 近似值;近似法;方法,途径 $vt.$ 趋近,接近
series $n.$ 级数,列
infinite series 无穷级数
figure $n.$ 图,图形;数字
sum $n.$ 和,总数
fundamental $n.$ 基本的
theorem $n.$ 定理
the Fundamental Theorem of Calculus 微积分学基本定理

2.2 THE DIFFERENTIAL

The problem of finding the tangent to a curve has been studied by many mathematicians since Archimedes explored the question in

Antiquity. The first attempt at determining the tangent to a curve that resembled the modern method of the Calculus came from Gilles Persone de Roberval during the 1630's and 1640's. At nearly the same time as Roberval was devising his method, Pierre de Fermat used the notion of maxima and the infinitesimal to find the tangent to a curve. Some credit Fermat with discovering the differential, but it was not until Leibniz and Newton rigorously defined their method of tangents that a generalized technique became accepted.

Roberval's Method of Tangent Lines Using Instantaneous Motion

The primary idea behind Roberval's method of determining the tangent to a curve was the notion of Instantaneous Motion. That is, he considered a curve to be sketched by a moving point. If, at any point on a curve, the vectors making up the motion could be determined, then the tangent was simply the combination (sum) of those vectors.

Roberval applied this method to find the tangents to curves for which he was able to determine the constituent motion vectors at a point. For a parabola, Roberval was able to determine such motion vectors.

Figure 8 depicts the graph of a parabola showing the constituent motion vectors V_1 and V_2 at a point P.

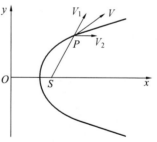

Figure 8

Roberval determined that at a point P in a parabola, there are two vectors accounting for its instantaneous motion. The vector V_1, which is in the same direction as the line joining the focus of the parabola (point S) and the point on the parabola (point P). The other vector making up the instantaneous motion (V_2) is perpendicular to the y-axis (which is the directrix, or the line perpendicular to the line bisecting the parabola). The tangent to the graph at point P is simply the vector sum $V = V_1 + V_2$.

Using this methodology, Roberval was able to find the tangents to numerous other curves including the ellipse and cycloid. However, finding the vectors describing the instantaneous motion at a point proved difficult for a large number of curves. Roberval was never able to generalize this method, and therefore exists historically only as a precursor to the method of finding tangents using infinitesimals.

Fermat's Maxima and Tangent

Pierre De Fermat's method for finding a tangent was developed during the 1630's, and though never rigorously formulated, is almost exactly the method used by Newton and Leibniz. Lacking a formal concept of a limit, Fermat was unable to properly justify his work. However, by examining his techniques, it is obvious that he understood precisely the method used in differentiation today.

In order to understand Fermat's method, it is first necessary to consider his technique for finding maxima. Fermat's first documented problem in differentiation involved finding the maxima of an equation, and it is clearly this work that led to his technique for finding tangents.

The problem Fermat considered was dividing a line segment into two segments such that the product of the two new segments was a maximum. In Figure 9, a line segment of length a is divided into two segments. Those two segments are x and $(a - x)$. Fermat's goal, then, was to maximize the product $x(a - x)$. His approach was mysterious at the time, but with the benefit of the current knowledge of limits, Fermat's method is quite simple to understand. What Fermat did was to replace each occurrence of x with $x + E$ and stated that when the maximum is found, x and $x + E$ will be equal. Therefore, he had the

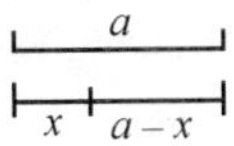

Figure 9

equation
$$x(a - x) = (x + E)(a - x - E)$$

Through simplifying both sides of the equation and canceling like terms, Fermat reduced it
$$E^2 - aE + 2xE = 0$$
$$E(E - a + 2x) = 0 \qquad E - a + 2x = 0$$

At this point, Fermat said to simply let $E = 0$, and as such one is left with
$$x = \left(\frac{a}{2}\right)$$

This says that to maximize the product of the two lengths, each length should be half the total length of the line segment. Though this result is correct, Fermat's method contains mysterious holes that are only rectified by current knowledge. Fermat simply lets $E = 0$, then in the step where he divides through by E, he would have division by zero. However, though Fermat formulated his method by saying $E = 0$, he was actually considering the limit of E as it *approaches zero* (which explains why his algebra works properly). Fermat's method of extrema can be understood in modern terms as well. By substituting $x + E$ for x, he is saying that $f(x + E) = f(x)$, or that $f(x + E) - f(x) = 0$. Since $f(x)$ is a polynomial, this expression will be divisible by E. Therefore, Fermat's method can be understood as the definition of the derivative (when used for finding extrema)
$$\lim_{E \to 0} \frac{f(x + E) - f(x)}{E} = 0$$

Although Fermat was never able to make a logically consistent formulation, his work can be interpreted as the definition of the differential.

Using his mysterious E, Fermat went on to develop a method for finding tangents to curves. Consider the graph of a parabola.

Fermat wishes to find a general formula for the tangent to $f(x)$. In order to do so, he draws the tangent line at a point x and will consider a point a distance E away. As can be seen from Figure 10, by similar triangles, the following relationship exists

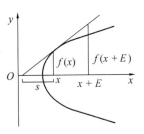

Figure 10

$$\frac{s}{s+E} = \frac{f(x)}{f(x+E)}$$

By isolating s, Fermat found that

$$s = \frac{f(x)}{\frac{f(x+E) - f(x)}{E}}$$

Fermat again lets the quantity $E = 0$ (in modern term, he took the limit as E approached 0) and recognized that the bottom portion of the equation was identical to his differential in his method of mimina. Consequently, in order to find the slope of a curve, all he needed to do was find $\frac{f(x)}{s}$. For example, consider the equation $f(x) = x^3$

$$s = \frac{f(x)}{\frac{f(x+E) - f(x)}{E}} = \frac{x^3}{\frac{(x+E)^3 - x^3}{E}} = \frac{x^3}{3x^2 + 3xE + E^2}$$

Again, Fermat lets $E = 0$ and finds that

$$s = \frac{x^3}{3x^2} = \frac{x}{3}$$

Now, returning to the original equation

$$f'(x) = \frac{f(x+E) - f(x)}{E} = \frac{f(x)}{s} = \frac{x^3}{\frac{x}{3}} = 3x^2$$

Here the modern notation for the derivative $f'(x)$ is used, which Fermat recognized to be equal to $\frac{f(x+E) - f(x)}{E}$ when he let $E = 0$.

Using this method, Fermat was able to derive a general rule for the tangent to a function $y = x^n$ to be nx^{n-1}. As described in the Integration section, Fermat had now developed a general rule for polynomial differentiation and integration. However, he never managed to see the inverse relationship between the two operations, and the logical inconsistencies in his justification left his work fairly unrecognized. It was not until Newton and Leibniz that this formulation became possible.

Newton and Leibniz

Newton and Leibniz served to complete three major necessities in the development of the Calculus. First, though differentiation and integration techniques had already been researched, they were the first to explain an "algorithmic process" for each operation. Second, despite the fact that differentiation and integration had already been discovered by Fermat, Newton and Leibniz recognized their usefulness as a general process. That is, those before Newton and Leibniz had considered solutions to area and tangent problems as specific solutions to particular problems. No one before them recognized the usefulness of the Calculus as a general mathematical tool. Third, though a recognition of differentiation and integration being inverse processes had occurred in earlier work, Newton and Leibniz were the first to explicitly pronounce and rigorously prove it.

Newton and Leibniz both approached the Calculus with different notations and different methodologies. The two men spent the latter part of their life in a dispute over who was responsible for inventing the Calculus and accusing each other of plagiarism. Though the names Newton and Leibniz are associated with the invention of the Calculus, it is clear that the fundamental development had already been forged by others. Though generalizing the techniques and explicitly showing the Fundamental Theorem of Calculus was no small feat, the mathematics involved in their methods are similar to those who came before them. Sufficiently similar are their methods that the specifics of their methodologies are beyond the scope of this paper. In terms of their mathematics, it is only their demonstration of the Fundamental Theorem of Calculus that will be discussed.

The Elusive Inverses — the Integral and Differential

The notation of Leibniz most closely resembles that which is used in modern calculus and his approach to discovering the inverse relationship between the integral and differential will be examined. Though Newton independently arrived at the same conclusion, his path to discovery is slightly less accessible to the modern reader.

Leibniz defined the differential as being

$$\frac{dy}{dx} = \lim_{\Delta x \to 0} \frac{\Delta y}{\Delta x}$$

From the earlier works of Cavalieri, Leibniz was already familiar with the techniques of finding the area underneath a curve. Leibniz discovered the inverse relationship between the area and derivative by utilizing his definition of the differential.

Consider the graph of the equation $y = x^2 + 1$:

Leibniz's idea was to use his differential on the area-function of the graph. Consider adding a Δ(area) underneath the graph of the curve. The Δ(area) is defined by the lower rectangle $PQSR$ with area is $y(\Delta x)$ plus a fraction of the upper rectangle $SRTU$ whose area is simply Δx (Δy). In other words, Δ(area) lies somewhere in between $y(\Delta x)$ and the total enclosing rectangle $PQTU$ whose area is $(y + \Delta y)(\Delta x)$. Leibniz then considered the ratio Δ(area)$/\Delta x$ and saw that since the Δ

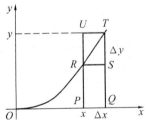

Figure 11

(area) is between $y(\Delta x)$ and $(y + \Delta y)(\Delta x)$ the ratio will be between y and $(y + \Delta y)$. From the diagram, it can be seen that Δx and Δy are closely related to each other. That is, as Δx approaches 0 so too does Δy. That means that the ratio $\Delta(\text{area})/\Delta x$ lies between y and a value that approaches y (since $y + \Delta y$ approaches y as Δy goes to 0). Written in terms of Leibniz's definition of the derivative

$$\frac{dy}{dx} = \lim_{\Delta x \to 0} \frac{\Delta \text{area}}{\Delta x} = y = x^2 + 1$$

Leibniz has shown the inverse relationship between the differential and the area-function. Namely that the differential of the area-function of a function y is equal to the function itself. In this case, the derivative of the area-function of $y = x^2 + 1$ is indeed $y = x^2 + 1$.

Leibniz's influence in the history of the integral spreads beyond finding this groundbreaking relationship. He was also responsible for inventing the notation that is used by most students of calculus today. Leibniz used the symbol $\int$ (which was simply how "S" was written at the time) to denote an infinite number of sums. This was closely related to what he called the "integral", or the sum of a number of infinitely small areas. The area underneath a function y, or integral of y, was expressed as $\int y\,(dx)$.

What Leibniz's notation was really saying was to sum up all of the areas $dx \times y$ as dx approached 0. As dx approaches 0, there are an infinite number of such areas, hence the symbolism $\int$ representing an infinite number of sums. Integration of this kind is also known as the *indefinite integral* or *anti-derivative* due to the inverse relationship found by Leibniz. That is, the derivative of the indefinite integral of a function yields the function itself. Leibniz also developed a notation for *definite integrals*, or integrals which produced the area underneath a curve between two bounding values (rather than a symbolic answer). His notation for the *definite integral* was to supply the lower and upper-bounding x-values with the integral symbol

$$\int_a^b f(x)\,dx = A(b) - A(a)$$

Where A is the area-function produced by the anti-derivative. The area function A was computed by using Wallis' law.

Words and Expressions

differential $n.$ 微分 $adj.$ 微分的

tangent $n.$ 正切,切线 $adj.$ 正切的,切线的

the tangent to a curve 曲线的切线

maxima $n.$ ($pl.$ form of maximum) 极大值,极大

tangent lines 切线

point $n.$ 点

a moving point 一个动点

any point on a curve 曲线上任意一点

focus $n.$ 焦点

the focus of the parabola 抛物线的焦点

perpendicular $n.$ 垂线 $adj.$ 垂直的,垂直于

y-axis y 轴

perpendicular to the y – axis 垂直于 y 轴

directrix $n.$ 准线

ellipse $n.$ 椭圆

cycloid $n.$ 摆线,旋轮线

differentiation $n.$ 微分法

line segment 线段

a line segment of length a 一个长度为 a 的线段

maximize $v.$ 使最大

equation $n.$ 方程

cancel $vt.$ 消去,消除; $vi.$ 抵消 $n.$ 约分

term *n*. 术语,项,条
like term 同类项
division *n*. 除法,除
extrema *n*. (*pl. form of extremum*) 极值
polynomial *n*. 多项式
derivative *n*. 导数,微商
inverse *n*. 逆
operations *n*. 运算
algorithmic *adj*. 算法的,规则系统的
process *n*. 方法,步骤;过程
algorithmic process 算法过程
indefinite *adj*. 不定的
the indefinite integral 不定积分
definite *adj*. 定的,确定的
the definite integral 定积分

2.3 THE CALCULUS

The main ideas which underpin the calculus developed over a very long period of time indeed. The first steps were taken by Greek mathematicians.

To the Greeks numbers were ratios of integers so the number line had "holes" in it. They got round this difficulty by using lengths, areas and volumes in addition to numbers for, to the Greeks, not all lengths were numbers.

Zeno of Elea, about 450 B.C., gave a number of problems which were based on the infinite. For example he argued that motion is impossible:

If a body moves from A to B then before it reaches B it passes through the mid-point, say B_1 of AB. Now to move to B_1 it must first reach the mid-point B_2 of AB_1. Continue this argument to see that A must move through an infinite number of distances and so cannot move.

Leucippus, Democritus and Antiphon all made contributions to the Greek method of exhaustion which was put on a scientific basis by Eudoxus about 370 B.C.. The method of exhaustion is so called because one thinks of the areas measured expanding so that they account for more and more of the required area.

However Archimedes, around 225 B.C., made one of the most significant of the Greek contributions. His first important advance was to show that the area of a segment of a parabola is 4/3 the area of a triangle with the same base and vertex and 2/3 of the area of the circumscribed parallelogram.

Archimedes constructed an infinite sequence of triangles starting with one of area A and continually adding further triangles between the existing ones and the parabola to get areas

$$A, A + \frac{A}{4}, A + \frac{A}{4} + \frac{A}{16}, A + \frac{A}{4} + \frac{A}{16} + \frac{A}{64}, \cdots$$

The area of the segment of the parabola is therefore

$$A\left(1 + \frac{1}{4} + \frac{1}{4^2} + \frac{1}{4^3} + \cdots\right) = \left(\frac{4}{3}\right) A$$

This is the first known example of the summation of an infinite series.

Archimedes used the method of exhaustion to find an approximation to the area of a circle. This, of course, is an early example of integration which led to approximate values of π.

Here is Archimedes' diagram

Among other "integrations" by Archimedes were the volume and surface area of a sphere, the volume and area of a cone, the surface area of an ellipse, the volume of any segment of a paraboloid of revolution and a segment of an hyperboloid of revolution.

No further progress was made until the 16th Century when mechanics began to drive mathematicians to examine problems such as centres of gravity. Luca Valerio (1552 ~ 1618) published *De quadratura parabolae* in Rome (1606) which continued the Greek methods of attacking these type of area problems. Kepler, in his work

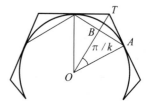

Figure 12

on planetary motion, had to find the area of sectors of an ellipse. His method consisted of thinking of areas as sums of lines, another crude form of integration, but Kepler had little time for Greek rigour and was rather lucky to obtain the correct answer after making two canceling errors in this work.

Three mathematicians, born within three years of each other, were the next to make major contributions. They were Fermat, Roberval and Cavalieri. Cavalieri was led to his "method of indivisibles" by Kepler's attempts at integration. He was not rigorous in his approach and it is hard to see clearly how he thought about his method. It appears that Cavalieri thought of an area as being made up of components which were lines and then summed his infinite number of "indivisibles". He showed, using these methods, that the integral of x^n from 0 to a was $\dfrac{a^{n+1}}{(n+1)}$ by showing the result for a number of values of n and inferring the general result.

Roberval considered problems of the same type but was much more rigorous than Cavalieri. Roberval looked at the area between a curve and a line as being made up of an infinite number of infinitely narrow rectangular strips. He applied this to the integral of x^m from 0 to 1 which he showed had approximate value

$$\frac{0^m + 1^m + 2^m + \cdots + (n-1)^m}{n^{m+1}}$$

Roberval then asserted that this tended to $\dfrac{1}{(m+1)}$ as n tends to infinity, so calculating the area.

Fermat was also more rigorous in his approach but gave no proofs. He generalized the parabola and hyperbola:

Parabola: $\dfrac{y}{a} = \left(\dfrac{x}{b}\right)^2$ to $\left(\dfrac{y}{a}\right)^2 = \left(\dfrac{x}{b}\right)^m$,

Hyperbola: $\dfrac{y}{a} = \left(\dfrac{b}{x}\right)^2$ to $\left(\dfrac{y}{a}\right)^n = \left(\dfrac{b}{x}\right)^m$.

In the course of examining $\dfrac{y}{a} = \left(\dfrac{x}{b}\right)^p$, Fermat computed the sum of r^p from $r = 1$ to $r = n$.

Fermat also investigated maxima and minima by considering when the tangent to the curve was parallel to the x-axis. He wrote to Descartes giving the method essentially as used today, namely finding maxima and minima by calculating when the derivative of the function was 0. In fact, because of this work, Lagrange stated clearly that he considers Fermat to be the inventor of the calculus.

Descartes produced an important method of determining normals in *La Géométrie* in 1637 based on double intersection. De Beaune extended his methods and applied it to tangents where double intersection translates into double roots. Hudde discovered a simpler method, known as Hudde's Rule, which basically involves the derivative. Descartes' method and Hudde's Rule were important in influencing Newton.

Huygens was critical of Cavalieri's proofs saying that what one needs is a proof which at least convinces one that a rigorous proof could be constructed. Huygens was a major influence on Leibniz and so played a considerable part in producing a more satisfactory approach to the calculus.

The next major step was provided by Torricelli and Barrow. Barrow gave a method of tangents to a curve where the tangent is given as the limit of a chord as the points approach each other known as *Barrow's differential triangle*.

Here is Barrow's differential triangle

Both Torricelli and Barrow considered the problem of motion with variable speed. The derivative of the distance is velocity and the inverse operation takes one from the velocity to the distance. Hence an awareness of the inverse of differentiation began

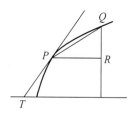

Figure 13

to evolve naturally and the idea that integral and derivative were inverses to each other were familiar to Barrow. In fact, although Barrow never explicitly stated the fundamental theorem of the calculus, he was working towards the result and Newton was to continue with this direction and state the Fundamental Theorem of the Calculus explicitly.

Torricelli's work was continued in Italy by Mengoli and Angeli.

Newton wrote a tract on fluxions in October 1666. This was a work which was not published at the time but seen by many mathematicians and had a major influence on the direction the calculus was to take. Newton thought of a particle tracing out a curve with two moving lines which were the coordinates. The horizontal velocity x' and the vertical velocity y' were the fluxions of x and y associated with the flux of time. The fluents or flowing quantities were x and y themselves. With this fluxion notation $\frac{y'}{x'}$ was the tangent to $f(x,y) = 0$.

In his 1666 tract Newton discusses the converse problem, given the relationship between x and $\frac{y'}{x'}$ find y. Hence the slope of the tangent was given for each x and when $\frac{y'}{x'} = f(x)$ then Newton solves the problem by antidifferentiation. He also calculated areas by antidifferentiation and this work contains the first clear statement of the Fundamental Theorem of the Calculus.

Newton had problems publishing his mathematical work. Barrow was in some way to blame for this since the publisher of Barrow's work had gone bankrupt and publishers were, after this, wary of publishing mathematical works! Newton's work on *Analysis with infinite series* was written in 1669 and circulated in manuscript. It was not published until 1711. Similarly his *Method of fluxions and infinite series* was written in 1671 and published in English translation in 1736. The Latin original was not published until much later.

In these two works Newton calculated the series expansion for $\sin x$ and $\cos x$ and the expansion for what was actually the exponential function, although this function was not established until Euler introduced the present notation e^x.

You can see the series expansions for sine and for cosine. They are now called Taylor or Maclaurin series.

Newton's next mathematical work was *Tractatus de Quadratura Curvarum* which he wrote in 1693 but it was not published until 1704 when he published it as an Appendix to his *Optiks*. This work contains another approach which involves taking limits. Newton says

In the time in which x by flowing becomes $x + \Delta x$, the quantity x^n becomes $(x + \Delta x)^n$

i.e. by the method of infinite series

$$x^n + n\Delta x x^{n-1} + \frac{(nn - n)}{2}\Delta x \Delta x x^{n-2} + \cdots$$

At the end he lets the increment Δx vanish by "taking limits". Leibniz learnt much on a European tour which led him to meet Huygens in Paris in 1672. He also met Hooke and Boyle in London in 1673 where he bought several mathematics books, including Barrow's works. Leibniz was to have a lengthy correspondence with Barrow. On returning to Paris

Leibniz did some very fine work on the calculus, thinking of the foundations very differently from Newton.

Newton considered variables changing with time. Leibniz thought of variables x, y as ranging over sequences of infinitely close values. He introduced dx and dy as differences between successive values of these sequences. Leibniz knew that $\dfrac{dy}{dx}$ gives the tangent but he did not use it as a defining property.

For Newton integration consisted of finding fluents for a given fluxion so the fact that integration and differentiation were inverses was implied. Leibniz used integration as a sum, in a rather similar way to Cavalieri. He was also happy to use "infinitesimals" dx and dy where Newton used x' and y' which were finite velocities. Of course neither Leibniz nor Newton thought in terms of functions, however, but both always thought in terms of graphs. For Newton the calculus was geometrical while Leibniz took it towards analysis.

Leibniz was very conscious that finding a good notation was of fundamental importance and thought a lot about it. Newton, on the other hand, wrote more for himself and, as a consequence, tended to use whatever notation he thought of on the day. Leibniz's notation of d and $\int$ highlighted the operator aspect which proved important in later developments. By 1675 Leibniz had settled on the notation

$$\int y \, dy = \frac{y^2}{2}$$

written exactly as it would be today. His results on the integral calculus were published in 1684 and 1686 under the name "calculus summatorius", the name integral calculus was suggested by Jacob Bernoulli in 1690.

After Newton and Leibniz the development of the calculus was continued by Jacob Bernoulli and Johann Bernoulli. However when Berkeley published his *Analyst* in 1734 attacking the lack of rigour in the calculus and disputing the logic on which it was based much effort was made to tighten the reasoning. Maclaurin attempted to put the calculus on a rigorous geometrical basis but the really satisfactory basis for the calculus had to wait for the work of Cauchy in the 19th Century.

Words and Expressions

volume $n.$ 体积,容积

exhaustion $n.$ 穷举,耗尽

method of exhaustion 穷举法

surface $n.$ 面,曲面

sphere $n.$ 球面,球形

surface area of a sphere 球的表面

paraboloid $n.$ 抛物面

revolution $n.$ 回转

paraboloid of revolution 回转抛物面

hyperboloid $n.$ 双曲面,双曲线体

hyperboloid of revolution 旋转双曲面

minima $n.$ ($pl.$ form of minimum) 极小值

variable $n.$ 变量,元,变元;变项

fluxions $n.$ 流数

exponential $n.$ 指数的

exponential function 指数函数

sequence $n.$ 序列

operator $n.$ 算子

2.4 SHORT BIOGRAPHIES

Introduction

The following chapter is meant to provide brief biographies of the mathematicians that significantly contributed to the development of the Calculus.

Gregory of St. Vincent (1584 ~ 1667)

A Jesuit teacher in Rome and Prague, he later became a tutor in the court of Philip IV of Spain. He tried to "square the circle"(constructing a square equal in area to circle using only a straight edge and compass) throughout his life, and discovered several interesting theorems while doing so. He discovered the expansion for $\log(1 + x)$ for ascending powers of x. Eventually, he thought he had squared the circle, but his method turned out to be equivalent to the modern method of integration. He successfully integrated x^{-1} in a geometric form which is equivalent to the natural logarithm function.

Rene Descartes (1596 ~ 1650)

Rene Descartes was a philosopher of great acclaim. The idea that humans may make mistakes in reasoning is the foundation of his philosophy. He cast aside all traditional beliefs and tried to build his philosophy from the ground up, based on his reasoning alone. In his search for a base on which he might begin to build his reconstructed view of the world, he doubted the reality of his own existence. The existence of his doubt persuaded him to formulate the famous maxim, "I think, therefore I am."

The precision and clarity of mathematics and mathematical reasoning impressed Descartes. He hoped to make use of it in the development of his philosophy and thereby reduce the susceptibility to flaws of his own reasoning. He spent a number of years studying mathematics and developing systematic methods for distinguishing between truth and falsehood. His contribution to the field of geometry can be thought of as an example of how his methods can be applied to reveal new truths, but to the mathematicians to follow him, Descartes' analytical geometry was powerful tool in its own right. Descartes' work in geometry laid the foundation for the calculus that was to come after him. Due to his wariness of mistakes in reasoning, Descres' tended to de-emphasize his formal education and instead focus on leaning by first-hand experience. His philosophy of experience led him to travel outside of his native France, serve in the military, and eventually live in Holland. During his time in Holland, Descartes tutored Princess Elisabeth , but devoted most of his time to contemplation of his philosophy and his writing. He was summoned to Sweden in 1646 to tutor Queen Christine, but the swedish winters were too difficult for him and Descartes died in 1650.

Bonaventura Cavalieri (1598 ~ 1647)

Cavalieri became a Jesuate (not a Jesuit as is frequently stated) at an early age it was because of that he was made a professor of mathematics at Boloagna in 1629. He held the position until his death in 1647. Cavalieri published tables of sines, tangents, secants, and versed sines along with their logarithms out to eight decimal places, but his most well known contribution is in the invention of the principle of indivisibles. His principle of indivisibles, developed by 1629, was first published in 1635 and was again published in 1653, after his death, with some corrections. The principle of indivisibles is based on the assumption that any line can be divided up into an infinite number of points, each having no length, a surface may be divided into an infinite number of lines, and a volume can be divided into an infinite number of surfaces.

Pierre de Fermat (1601 ~ 1665)

Pierre de Fermat was born in France, near Montauban, in 1601, died at Castres on January 12, 1665. Fermat was the son of a leather merchant, and he was educated at home. He became a councilor for the local parliament at Toulouse in 1631, a job where he spent the rest of his life. Fermat's life, except for a dispute with Descartes, was peaceful and unremarkable. The field of mathematics was a hobby for Fermat. He did not publish much during his lifetime regarding his findings. Some of his most important contributions to mathematics were found after his death, written in the margins of works he had read or contained within his notes. He did not seem to intend for any of his work to be published, for he rarely gave any proof with his notes of hid discoveries. Pierre de Fermat's interests were focused in there areas of mathematics: the theory of numbers, the use of geometry of analysis and infinitesimals, and probability. Math was a hobby for Fermat — his real job was as a judge. Judge of the day were expected to be aloof(so as to resist bribery), so he had a lot of time for his hobby.

Gilles Persone de Roberval (1602 ~ 1675)

Held chair of Ramus at the College for 40 years from 1634. He developed a method of indivisibles similar to that of Cavalieri, but did not disclose it. Roberval became involved in a number of disputes about priority and credit; the worst of these concerned cycloids. He developed a method find the area under a cycloid. Some of his more useful discoveries were computing the definite integral of $\sin x$, drawing the tangent to a curve, and computing the arc length of a spiral. Roberval called a cycloid a trochoid, which is Greek for wheel.

John Wallis (1616 ~ 1703)

A professor of geometry at Oxford, he had several very important publications, which advanced the field of indivisibles. He studied the work of Cavalieri, Descartes, Kepler, Rberval, and Torrcelli. He introduced ideas in calculus that went beyond those he read of. He discovered methods to evaluate integrals that were later used by Newton in his work on the binomial theorem. Wallis was the first to use the modern symbol for infinity. It is interesting to note that Wallis rejected the idea that negative numbers were less than nothing but accepted the notion that they were greater than infinity.

Blaise Pascal (1623 ~ 1662)

Pascal was a French student of Desargues. Etienne Pascal, Blaise's father, kept him away from mathematical texts early in his life until Blaise was twelve, when he studied geometry on his own. After this, Etienne, himself a mathematician, urged Blaise to study. Pascal invented an adding machine, to aid in his father's job as a tax collector. Its development was hindered by the units of currency used in France and England at the time. Two hundered and forty deniers equaled one liver, which is a difficult ratio for conversion.

Pascal turned to religion at the age of twenty-seven, ceasing to work on any mathematical problems. When he had to administer his father's estate for a time, he returned to studying the pressure of gasses and liquids, which got him into many arguments because he believed that there is a vacuum above the atmosphere; an unpopular belief at the time. During this time he also founded the theory of probability with Fermat. Late in his life, he turned to the study of the cycloid when he had a toothache. The toothache went away immediately upon pondering a cycloid, and he took this as a sign to study more on the subject of cycloids.

In Pascal's *Pensées*, one of his large religious papers, Pascal made a famous statement known as Pascal's Wager: "If God does not exist, one will lose nothing by believing in him, while if he does exist, one will lose everthing by not believing." His conclusion was that "... we are compelled to gamble..."

Christiaan Huygens (1629 ~ 1695)

Descartes took interest in Huygens at an early age and influenced his methematical education. He developed new mathods of grinding and polishing telescope lenses, and using a lens he made, he was able to see the first moon of Saturn. He was the one to discover the shape of the rings around Saturn using his improved telescopes. Huygens patented the first pendulum clock, which was able to keep more accurate time than current clocks because he need a way to keep more accurate time for his astronomical observations. He was elected to the Royal Society of London and also to the *Académie Royale des Sciences* in France. Leibniz was a frequent visitor to the *Académie* and learned much of his mathematics from Huygens. Throughout his life he worked on pendulum clocks to determine longitude at sea.

In one of his books, he describes the descent of heavy bodies in a vacuum in which he shows that the *cycloid* is the *tautochrone*, which means it is the shortest path. He also shows that the force on a body moving in a circle of radius r with a constant velocity of v varies directly as v^2 and inversely as r.

Isaac Barrow (1630 ~ 1677)

An Englishman, he was ordained and later made a professor of geometry at Gresham College in London. Barrow developed a method for determining tangents that closely approached the methods of calculus. He was also the first to discover that differentiation and integration were inverse operations. He thought that algebra should be part of logic instead of mathematics, which hindered his search for analytic discoveries. Barrow published a method for finding tangents, which turned out to be an improvement on Fermat's method of tangents. He worked with Cavalieri, Huygens, Gregory of St. Vincent, James Gregory, Wallis, and Newton.

James Gregory (1638 ~ 1675)

A Scotsman, he was familiar with the mathematics of several countries. Gregory worked with infinite series expansion, and infinite processes in general. He sought to prove, through infinite processes, that one could not square the circle, but Huygens, who was regarded as the leading mathematician of the day, believed that pi could be expressed algebraically, and many questioned the validity of Gregory's methods. Two hundred years later, it was proved that Gregory was right.

Much of his work was expressed in geometric terms, which was more difficult to follow than if it has been expressed algebraically. Because of this, Newton was the first to invent Calculus, even though Gregory knew all the important elements of Calculus, they were not expressed in a form that was easily understandable. Gregory only has the infinite series for arctangent attributed to him, even though he also discovered the infinite series for tangent, arcsecant, cosine, arccosine, sine, and arcsine. Using his infinite series for arctangent, he was able to find an expansion for $\pi/4$ several years before Leibniz.

Sir Isaac Newton (1642 ~ 1727)

Newton's father was a farmer, and it was intended that he follow in the family business. Instead of running the farm, an uncle decided that he should attend college, specifically Trinity College in Cambridge, where the uncle has attended college. Newton's original objective was to obtain a law degree. He attended Barrow's lectures and originally studied geometry only as a means to understand astronomy. In 1665, Trinity College closed down because of the plague in England. During the year it was closed he made several important discoveries. He developed the foundation for his integral and differential calculus, his universal theory of gravitation and also some theories about color. Upon his return to Trinity, Barrow resigned the Lucasian chair in 1669, and recommended Newton for the position. He continued to work on optics and mathematical problems until 1693, when he had a nervous breakdown. He took up a government position in London, ceasing all research. In 1708 , Newton was knighted by Queen Anne; he was honored for all his scientific work. He was elected president to the Royal Society in 1703 and held the position until his death in 1727.

Leibniz (1646 ~ 1716)

Leibinz was a law student at the University of Leipzig and the University of Altdorf. In 1672, he traveled to Paris in order to try and dissuade Louis XIV from attacking German areas. He stayed in Paris until 1676. During this time he continaed to study law but also studied physics and mathematics under Huygens. During this time he developed the basic version of his calculus. In one of his manuscripts dated November 21, 1675, he used the current-day notation for the integral, and also gave the product rule for differentiation. By 1676, he had discovered the power rule for both integral and fractional powers. In 1684 he published a paper containing the now-common notation, the rules for computing derivatives of powers, products and quotients. One year later, Newton published his Principia. Because Newton's work was published after leibniz's, there was a great dispute over who discovered the theories of calculus first which went on past their deaths.

Leonhard Euler (1707 ~ 1783)

Euler was the son of a Lutheran minister and was educated in his native town under the direction of John Bernoulli. He formed a life-long friendship with John Bernoulli's sons, Daniel and Nicholas. Euler went to the St. Petersburg Academy of Science in Russia with Daniel Bernoulli at the invitation of the empress. The harsh climate in Russia affected his eyesight; he lost the use of one eye completely in 1735. In 1741, Euler moved to Berlin at the command of Frederick the Great. While in Berlin, he wrote over 200 articles and three books on mathematical analysis. Euler did not get along well with Frederick the Great, however, and he returned to Russia in 1766. Within three years, he had become totally blind. Even though he was blind, he continued his work and published even more works. Euler produced a total of 886 books and papers through his life. After he died, the St. Petersburg Academy continued to publish his unpublished papers for 50 years. Euler used the notations $f(x)$, i for the square root of -1, π for pi, $\sum$ for summation, and e for the base of a natural logarithm. Euler died in 1783 of apoplexy.

Words and Expressions

significantly *adv.* 值得注目地,非偶然地;另有含义地
contribute *vt.* 贡献,捐助
significantly contributed 值得注目地贡献
Prague 布拉格(捷克首都)
tutor *n.* (英国大学)导师;(美国大学)教员(职位低于讲师)
construct *vt.* 组成,构造,设计;作(圆)
compass *n.* 圆规,指南针,范围
ascending *adj.* 上升的;向上的;增长的
ascending power 升幂
the natural logarithm function 自然对数

precision *n*. 精度,精确度,精确
clarity *n*. 明晰
susceptibility *n*. 感受性
flaws *vt*. 使有裂痕,使有瑕疵 *vi*. 有裂痕,有瑕疵 *n*. 裂缝,裂纹;缺陷,瑕疵
systematic *adj*. 系统的,有系统的
systematic method 系统法
reveal *vt*. 泄露,透露;显示,显出
analytical *adj*. 解析的,分析的
geometry *n*. 几何,几何学
analytical geometry 解析几何
summon *vt*. 召唤,传唤;召集
secant *n*. 正割
versed sine 反正弦
leather *n*. 皮革;皮革制品 *adj*. 皮革制的,皮的
leather merchant 皮革商人
theory of numbers 数论
aloof *adv*. 远离,躲开 *adj*. 冷漠的,疏远的,无情的
bribery *n*. 行贿或受贿的行为
priority *n*. 优先权
arc length 弧长
spiral *n*. 螺线,蜷线
cease *vi*. 停止;终止
estate *n*. 地产,财产
vacuum *n*. 真空;真空吸尘器 *adj*. 真空的;产生真空的
longitude *n*. 经度
ordain *vi*. 注定,任命 *vt*. 注定,规定;任命
Gresham College 葛莱兴学院
arcsecant *n*. 反正割
arccosine *n*. 反余弦

arctangent *n.* 反正切
universal theory of gravitation 万有引力
optics *n.* 光学
version *n.* 译文,翻译
manuscript *n.* 手稿,草稿,原稿
quotient *n.* 商
Lutheran *n.* 路德教徒 *adj.* 马丁路德的,路德教派的
harsh *adj.* 粗糙的;严厉的,苛刻的;恶劣的
mathematical analysis 数学分析
notation *n.* 记号,记法
the St. Petersburg Academy 圣彼德堡科学院
apoplexy *n.* 脑溢血

The Algebra

3.1 NOTATIONS AND A REVIEW OF NUMBERS

The Language of Sets

The language of sets pervades all of mathematics. It provided a convenient shorthand for expressing mathematical statements. Loosely speaking, a set can be defined as a collection of objects, who called the *members* of the set. This definition will suffice for us. We use some shorthand to indicate certain relationships between sets and elements. Usually, sets will be designated by upper case letters such as A, B, etc., and elements will be designated by lower case letters such as a, b, etc. As usual, a set A is a subset of the set B if every element of A is an element of B, and a proper subset if it is a subset not equal to B. Two sets A and B are said to be *equal* if they have exactly the same elements. Some shorthand:

Set Operations

ϕ denotes the empty set, i.e., the set with no members.

$a \in A$ means "a is a member of the set A".

$A = B$ means "the set A is equal to the set B".

$A \subset B$ means "A is a subset of B".

$A \subsetneq B$ means "A is a proper subset of B".

There are two ways in which we may prescribe a set: we may *list* its elements, such as in the definition $A = \{0,1,2,3\}$ or specify them by *rule* such as in the definition $A = \{x \mid x \text{ is an integer and } 0 \leq x \leq 3\}$. (Read this as A is the set of x such that x is an integer and $0 \leq x \leq 3$.) With this notation we can give formal definitions of set intersections and unions:

Definition 1 Let A and B be sets. Then the *intersection* of A and B is defined to be the set $A \cap B = \{x \mid x \in A \text{ and } x \in B\}$. The *union* of A and B is the set $A \cup B = \{x \mid x \in A \text{ or } x \in B\}$. The *difference* of A and B is the set $A - B = \{x \mid x \in A \text{ and } x \notin B\}$.

Example 1 Let $A = \{0,1,3\}$ and $B = \{0,1,2,4\}$. Then

$A \cup \phi = A$

$A \cap \phi = \phi$

$A \cup B = \{0,1,2,3,4\}$

$A \cap B = \{0,1\}$

$A - B = \{3\}$

About Numbers

One could spend a full course fully developing the properties of number systems. We won't do that, of course, but we will review some of the basic sets of the numbers, and assume the reader is familiars with properties of numbers we have not mentioned here. At the start of it all are the kind of numbers that every child knows something about the

natural or *counting* numbers. This is the set
$$\mathbf{N} = \{1, 2, 3, \cdots\}$$
One could view most subsequent expansions of the concept of number as a matter of rising to the challenge of solving equations. For example, we cannot solve the equation
$$x + m = n, \quad m, n \in \mathbf{N}$$
for the unknown x without introducing subtraction and extending the notion of natural number that of *integer*. The set of integers is denoted by
$$\mathbf{Z} = \{0, \pm 1, \pm 2, \cdots\}$$
Next, we cannot solve the equation
$$ax = b, \quad a, b \in \mathbf{Z}$$
for the unknown x with introducing division and extending the notion of integer to that *rational number*. The set of rationals is denoted by
$$\mathbf{Q} = \left\{ \frac{a}{b} \mid A, B \in \mathbf{Z} \text{ and } b \neq 0 \right\}$$
Rational number arithmetic has some characteristics that distinguish it from integer arithmetic. The main difference is that nonzero rational numbers have multiplicative inverses (the multiplicative inverse of $\frac{a}{b}$ is $\frac{b}{a}$). Such a number system is called a *field* of numbers. In a nutshell, a *field of numbers* is a system of objects, called numbers, together with operations of addition, subtraction, multiplication and division that satisfy the usual arithmetic laws: in particular, it must be possible to subtract any number from any other and divide any number by a nonzero number to obtain another such number. The associative, commutative, identity and inverse laws must hold for each of addition and mutiplication: and the distributive law must hold for multiplication over addition. The rationals form a field of numbers: the integers don't since division by nonzero integers is not always possible if we restrict our numbers to integers.

The jump from rational to real numbers cannot be entirely explained

by algebra, although algebra offers some insight as to why the number system still needs to be extended. An equation like
$$x^2 = 2$$
does not have a rational solution, since $\sqrt{2}$ is irrational. (Story has it that this is lethal knowledge, in that followers of a Pythagorean cult claim that the gods threw overboard a ship one of their followers who was unfortunate enough to discover the fact.) There is also the problem of numbers like π and Euler's constant e which do not even satisfy any polynomial equation. The heart of the problem is that if we only consider rationals on a number line, there are many "holes" which are filled by number like π or $\sqrt{2}$. Filling in these holes leads us to the set **R** of real numbers, which are in one-to-one correspondence with the points on a number line. We won't give an exact definition of the set of real numbers. Recall that every real number admits a (possibly infinite) decimal representation, such as $\dfrac{1}{3} = 0.333\cdots$ or $\pi = 3.141\,59\cdots$. This provides us with a loose definition: real numbers are numbers that can be expressed by a decimal representation, i.e., limits of finite decimal expansions.

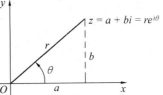

Figure 14

There is one more problem to overcome. How do we solve a system like
$$x^2 + 1 = 0$$
over the reals? The answer is we can't: if x is real, then $x^2 \geqslant 0$, so $x^2 + 1 > 0$. We need to extend our number system one more time, and this leads to the set **C** of *complex* number. We define i to be a quantity such that $i^2 = -1$ and
$$\mathbf{C} = \{a + bi \mid a, b \in \mathbf{R}\}$$
If the complex number $z = a + bi$ is given, then we say that the

form $a + b\mathrm{i}$ is the *standard form* for z. In this case the real part of z is $\mathbf{R}(z) = a$ and the imaginary part is defined as $\mathbf{I}(z) = b$. (Notice that the imaginary part of z is a real number: it is the real coefficient of i.) Two complex numbers are equal precisely when they have the same real parts and the same imaginary parts. All of this could be put on a more formal basis by initially defining complex numbers to be ordered pairs of real numbers. We will not do so, but the fact that complex numbers behave like ordered pairs of real numbers leads to an important geometrical insight: complex numbers can be identified with points in the plane. Instead of an x and y axis, one lays out a *real* and *imaginary* axis (which is still usually labeled with x and y) and plots complex numbers $a + b\mathrm{i}$ as in Figure 14. This results in the so-called *complex plane*.

Arithmetic in $\mathbf{C}$ is carried out by using the usual laws of arithmetic for $\mathbf{R}$ and the algebraic identity $\mathrm{i}^2 = -1$ to reduce the result to standard form. Thus we have the following laws of complex arithmetic

$$(a + b\mathrm{i}) + (c + d\mathrm{i}) = (a + c) + (b + d)\mathrm{i}$$
$$(a + b\mathrm{i}) \cdot (c + d\mathrm{i}) = (ac - bd) + (ad + bc)\mathrm{i}$$

In particular, notice that complex addition is exactly like the vector addition of plane vectors. Complex multiplication does not admit such a simple interpretation.

Example 2   Let $z_1 = 2 + 4\mathrm{i}$ and $z_2 = 1 - 3\mathrm{i}$. Compute $z_1 - 3z_2$.
Solution   We have that

$$z_1 - 3z_2 = (2 + 4\mathrm{i}) - 3(1 - 3\mathrm{i}) = 2 + 4\mathrm{i} - 3 + 9\mathrm{i} = -1 + 13\mathrm{i}$$

There are several more useful ideas about complex numbers that we will need. The *length* or *absolute value* of a complex number $z = a + b\mathrm{i}$ is defined as the nonnegative real number $|z| = \sqrt{a^2 + b^2}$, which is exactly the length of z viewed as a plane vector. The *complex conjugate* of z is defined as $\overline{z} = a - b\mathrm{i}$. Some easily checked and very useful facts about absolute value and complex conjugation

$$|z_1 z_2| = |z_1||z_2|$$
$$|z_1 + z_2| \leq |z_1| + |z_2|$$
$$|z|^2 = z\bar{z}$$
$$\overline{z_1 + z_2} = \bar{z_1} + \bar{z_2}$$
$$\overline{z_1 z_2} = \bar{z_1}\,\bar{z_2}$$

Example 3 Let $z_1 = 2 + 4i$ and $z_2 = 1 - 3i$. Verify for this z_1 and z_2 that $|z_1 z_2| = |z_1||z_2|$.

Solution First calculate that $z_1 z_2 = (2 + 4i)(1 - 3i) = (2 + 12) + (4 - 6)i$ so that $|z_1 z_2| = \sqrt{14^2 + (-2)^2} = \sqrt{200}$, while $z_1 = \sqrt{2^2 + 4^2} = \sqrt{20}$ and $|z_2| = \sqrt{1^2 + (-3)^2} = \sqrt{10}$. It follows that $|z_1 z_2| = \sqrt{10}\sqrt{20} = |z_1||z_2|$.

Example 4 Verify that the last fact in the product of conjugates is the product.

Solution This is just the last fact in the preceding list. Let $z_1 = x_1 + iy_1$ and $z_2 = x_2 + iy_2$ be in standard form, so that $\bar{z_1} = x_1 - iy_1$ and $\bar{z_2} = x_2 - iy_2$. We calculate

$$z_1 z_2 = (x_1 x_2 - y_1 y_2) + i(x_1 y_2 + x_2 y_1)$$

so that

$$\overline{z_1 z_2} = (x_1 x_2 - y_1 y_2) - i(x_1 y_2 + x_2 y_1)$$

Also

$$\bar{z_1}\,\bar{z_2} = (x_1 - iy_1)(x_2 - iy_2) = (x_1 x_2 - y_1 y_2) + (-i(x_1 y_2 + x_2 y_1)) = \overline{z_1 z_2}$$

The complex number i solves the equation $x^2 + 1 = 0$ (no surprise here: it was invented expressly for that purpose). The big surprise is that once we have the complex numbers in hand, we have a number system so complete that we can solve any polynomial equation in it. We won't offer a proof of this fact — it's very nontrivial. Suffice it to say that nineteenth century mathematicians considered this fact so fundamental that they dubbed it the "Fundamental Theorem of Algebra," a terminology we adopt.

Theorem 1 Let $p(z) = a_n z^n + a_{n-1} z^{n-1} + \cdots + a_1 z + a_0$ be a non-constant polynomial in the variable z with complex coefficients $a_0, \cdots, a_n$. Then the polynomial equation $p(z) = 0$ has a solution in the field C of complex numbers.

Note that the Fundamental Theorem doesn't tell us how to find a root of a polynomial — only that it can be done. As a matter of fact, there are no general formulas for the roots of a polynomial of degree greater than four, which means that we have to resort to numerical approximations in most cases.

In vector space theory the numbers in use are sometimes called *scalars*, and we will use this term. Unless otherwise stated or suggested by the presence of i, the field of scalars in which we do arithmetic is assumed to be the field of real numbers. However, we shall see later when we study eigensystems, that even if we are only interested in real scalars, complex numbers have a way of turning up quite naturally.

Let's do a few more examples of complex number manipulation.

Example 5 Solve the linear equation $(1 - 2i)z = (2 + 4i)$ for the complex variable z. Also compute the complex conjugate and absolute value of the solution.

Solution The solution requires that we put the complex number $z = \dfrac{2 + 4i}{1 - 2i}$ in standard form. Proceed as follows: multiply both numerator and denominator by $\overline{(1 - 2i)} = 1 + 2i$ to obtain that

$$z = \frac{2 + 4i}{1 - 2i} = \frac{(2 + 4i)(1 + 2i)}{(1 - 2i)(1 + 2i)} = \frac{2 - 8 + (4 + 4)i}{1 + 4} = \frac{-6}{5} + \frac{8}{5}i$$

Next we see that

$$\bar{z} = \overline{\frac{-6}{5} + \frac{8}{5}i} = \frac{6}{5} - \frac{8}{5}i$$

and

$$|z| = \left|\frac{1}{5}(-6+8i)\right| = \frac{1}{5}|(-6+8i)| =$$
$$\frac{1}{5}\sqrt{(-6)^2+8^2} = \frac{10}{5} = 2$$

Words and Expressions

set *n*. 集,集合
member *n*. 元
members of the set 集合的元
element *n*. 元,元素
empty *adj*. 空
empty set 空集
subset *n*. 子集
proper *adj*. 真,正常,常态,固有的
proper subset 真子集
intersections *n*. 交,相交
unions *n*. 并,并集
complex *n*. 复形 *adj*. 复的
complex number 复数
plane *n*. 平面
imaginary *adj*. 虚的
algebra *n*. 代数,代数学
the Fundamental Theorem of Algebra 代数基本定理
root *n*. 根
scalar *n*. 标量,数量
eigensystem *n*. 本征系统
linear equation 线性方程

3.2 LINEAR ALGEBRA

"Linear algebra" is the study of linear sets of equations and their transformation properties. Linear algebra allows the analysis of rotations in space, least squares fitting, solution of coupled differential equations, determination of a circle passing through three given points, as well as many other problems in mathematics, physics, and engineering.

The matrix and determinant are extremely useful tools of linear algebra. One central problem of linear algebra is the solution of the matrix equation

$$Ax = b$$

for x. While this can, in theory, be solved using a matrix inverse

$$x = A^{-1} b$$

other techniques such as Gaussian elimination are numerically more robust.

In addition to being used to describe the study of linear sets of equations, the term "linear algebra" is also used to describe a particular type of algebra. In particular, a linear algebra L over a field F has the structure of a ring with all the usual axioms for an inner addition and an inner multiplication together with distributive laws, therefore giving it more structure than a ring. A linear algebra also admits an outer operation of multiplication by scalars (that are elements of the underlying field F). For example, the set of all linear transformations from a vector space V to itself over a field F forms a linear algebra over F. Another example of a linear algebra is the set of all real square matrices over the field of the real numbers.

Linear Systems of Equations

The problem of finding all solutions to a linear system and that of finding an eigensystem for a square matrix is one central problem about which much of the theory of linear algbra. It requires some background development and even the motivation for this problem is fairly sophisticated. And it is easy to understand and motivate. As a matter of fact, simple cases of this problem are a part of the high school algebra background of most of us. We will address the problem of when a linear system has a solution and how to solve such a system for all of its solutions. Examples of linear system appear in nearly every scientific discipline: we touch on a few in the following.

Some Examples

Here are a few elementary examples of linear systems:

Example 1 For what values of the unknowns x and y are the following equations satisfied?

$$\begin{cases} x + 2y = 5 \\ 4x + y = 6 \end{cases}$$

Solution The first way that we taught to solve this problem was the geometrical approach: every equation of the form $ax + by + c = 0$ represents the graph of a straight line, and conversely, every line in the xy - plane is so described. Thus, each equation above represents a line. We need only graph each of the lines, then look for the point where these lines intersect, to find the unique solution to the graph (see Figure 15). Of course, the two equations may represent the same line, in which case there are infinitely many solutions, or distinct parallel lines, in which case there are no solutions. These could be viewed as exceptional or "degenerate" cases. Normally, we expect the solution to be unique, which it is in this example.

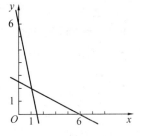

Figure 15

We also learned how to solve such an equation algebraically: in the present case we may use either equation to solve for one variable, say x, and substitute the result into the other equation to obtain an equation which is solved for y. For example, the first equation above yields $x = 5 - 2y$ and substitute into the second yields $4(5 - 2y) + y = 6$, i.e., $-7y = -14$, so that $y = 2$. Now substitute 2 for y in the first equation and obtain that $x = 5 - 2(2) = 1$.

Example 2 For what values of the unknowns x, y and z are the following equations satisfied

$$\begin{cases} x + y + z = 4 \\ 2x + 2y + 5z = 11 \\ 4x + 6y + 8z = 24 \end{cases}$$

Solution The geometrical approach becomes somewhat impractical as a means of obtaining an explicit solution to our problem: graphing in three dimensions on a flat sheet of paper doesn't lead to very accurate answers! Nonetheless, the geometrical point of view is useful, for it gives us an idea of what to expect without actually solving the system of equations.

With reference to our system of three equations in three unknowns, the first fact to take note of is that each of the three equations is an instance of the general equation $ax + by + cz + d = 0$. Now we know from analytical geometry that the graph of this equation is a plane in three dimensions, and conversely every such plane is the graph of some equation of the above form. In general, two planes will intersect in a line, though there are exceptional cases of the two planes represented being identical or distinct and parallel. Hence we know the geometrical shape of the solution set to the first two equations of our system: a plane, line or point. Similarly, a line and plane will intersect in a point or, in the exceptional case that the line and plane are parallel, their intersection will be the line itself or the empty set. Hence, we know that the above system of three equations has a solution set that is either empty, a single point, a line or a plane.

Which outcome occurs with our system of equations? We need the algebraic point of view to help us caculate the solution. The matter of dealing with three equations and three unknowns is a bit trickier than the problem of two equations and unknowns. Just as with two equations and unknowns, the key idea is still to use one equation to solve for one unknown. Since we have used one equation up, what remains is two equations in the remaining unknowns. In this problem, subtract 2 times the first equation from the second and 4 times the first equation from the third to obtain the system

$$\begin{cases} 3z = 3 \\ 2y + 4z = 8 \end{cases}$$

Which are easily solved to obtain $z = 1$ and $y = 2$. Now substitute into the first equation and obtain that $x = 1$. We can see that the graphical method of solution becomes impractical for systems of more than two variables, though it still tells us about the qualitative nature of the solution. This solution can be discerned roughly in Figure 16.

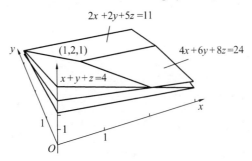

Figure 16

Some Key Notation

Here is a formal statement of the kind of equation. This formulation gives us a means of dealing with the general problem later on.

Definition 1 A *linear equation* in the variables $x_1, x_2, \cdots, x_n$ is an equation of the form

$$a_1 x_1 + a_2 x_2 + \cdots + a_n x_n = b$$

where the coefficients $a_1, a_2, \cdots, a_n$ and right hand side constant term b are given constant.

Of course, there are interesting and useful nonlinear equations, such as $ax^2 + bx + c = 0$, or $x^2 + y^2 = 1$, etc. But our focus is on systems that consist solely of linear equations. In fact, our next definition gives a fancy way of describing the general linear system.

Definition 2 A *linear system* of m equations in the n unknowns $x_1, x_2, \cdots, x_n$ is a list of m equations of the form

$$\begin{cases} a_{11} x_1 + a_{12} x_2 + \cdots + a_{1j} x_j + \cdots + a_{1n} x_n = b_1 \\ a_{21} x_1 + a_{22} x_2 + \cdots + a_{2j} x_j + \cdots + a_{2n} x_n = b_2 \\ \quad \vdots \qquad\qquad\qquad\qquad\quad \vdots \quad\ \vdots \\ a_{i1} x_1 + a_{i2} x_2 + \cdots + a_{ij} x_j + \cdots + a_{in} x_n = b_i \\ \quad \vdots \qquad\qquad\qquad\qquad\quad \vdots \quad\ \vdots \\ a_{m1} x_1 + a_{m2} x_2 + \cdots a_{mj} x_j + \cdots + a_{mn} x_n = b_m \end{cases} \qquad (1)$$

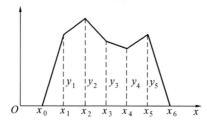

Figure 17

Notice how the coefficients are indexed: in the ith row the coefficient of the jth variable, x_j, is the number a_{ij}, and the right hand side of the ith equation is b_i. This systematic way of describing the system will come in handy later, when we introduce the matrix concept.

Examples of Modeling Problems

It is easy to get the impression that linear algebra is about the simple kinds of problems of the preceding examples. So why develop a whole subject? Next we consider two examples whose solutions will not be so apparent as the previous two examples. The real point of this part is to develop algebraic and geometrical methodologies which are powerful enough to handle problems like these.

We consider a diffusion process arising from the flow of heat through a homogeneous material substance. A basic physical observation to begin with is that heat is directly proportional to temperature. In a wide range of problems this hypothesis is ture, and we shall always assume that we are modeling such a problem. Thus, we can measure the amount of heat at a point by measuring temperature since they differ by a known constant of proportionality. To fix ideas, suppose we have a rod of material of unit length, say, situated on the x-axis, for $0 \leqslant x \leqslant 1$. Suppose further that the rod is laterally insulated, but has a known internal heat source that doesn't change with time. When sufficient time passes, the temperature of the rod at each point will settle down to "steady state" values, dependent only on position x. Say the heat source is described by a function $f(x)$, $0 \leqslant x \leqslant 1$, which gives the additional temperature contribution per unit length per unit time due to the heat source at the point x. Also suppose that the left and right ends of the rod are held at fixed at temperatures y_0 and y_1.

How can we model a steady state? Imagine that the continuous rod of uniform materials divided up into a finite number of equally spaced points, called nodes, namely $x_0 = 0, x_1, \cdots, x_{n+1} = 1$ and that all the heat is concentrated at these points. Assume the nodes are a distance h apart. Since spacing is equal, the relation between h and n is $h =$

$\frac{1}{n+1}$. Let the temperature function be $y(x)$ and let $y_i = y(x_i)$. Approximate $y(x)$ in between nodes by connecting adjacent points (x_i, y_i) with a line segment. (See Figure 17 for a graph of the resulting approximation to $y(x)$.) We know that at the end nodes the temperature is specified: $y(x_0) = y_0$ and $y(x_{n+1}) = y_1$. By examining the process at each interior node, we can obtain the following linear equation for each interior node index $i = 1, 2, \cdots, n$ involving a constant k called the conductivity of the material. A derivation of these equations follows this example

$$k \frac{-y_{i-1} + 2y_i - y_{i+1}}{h^2} = f(x_i)$$

or
$$-y_{i-1} + 2y_i - y_{i+1} = \frac{h^2}{k} f(x_i) \tag{2}$$

Example 3 Suppose we have a rod of material of conductivity $k = 1$ and situated on the x-axis, for $0 \leq x \leq 1$. Suppose further that the rod is laterally insulated, but has a known internal heat source and that both the left and right ends of the rod are held at 0 degrees Fathrenheit. What are the steady state equations approximately for this problem?

Solution Follow the notation of the discussion preceding this example. Notice that in this case $x_i = ih$. Remember that y_0 and y_{n+1} are known to be 0, so the terms y_0 and y_{n+1} disappear. Thus we have from equation 2 that there are n equations in the unknowns y_i, $i = 1, 2, \cdots, n$.

It is reasonable to expect that the smaller h is, the more accurately y_i will approximate $y(x_i)$. This is indeed the case. But consider what we are confronted with when we take $n = 5$, i.e., $h = \frac{1}{5+1} = \frac{1}{6}$, which is hardly a small value of h. The system of five equations in five unknowns becomes

$$\begin{cases} 2y_1 - y_2 &= \dfrac{f\left(\frac{1}{6}\right)}{36} \\ -y_1 + 2y_2 - y_3 &= \dfrac{f\left(\frac{2}{6}\right)}{36} \\ -y_2 + 2y_3 - y_4 &= \dfrac{f\left(\frac{3}{6}\right)}{36} \\ -y_3 + 2y_4 - y_5 &= \dfrac{f\left(\frac{4}{6}\right)}{36} \\ -y_4 + 2y_5 &= \dfrac{f\left(\frac{5}{6}\right)}{36} \end{cases}$$

This problem is already about as large as we would want to work by hand. The basic ideas of solving systems like this are the same as in Example 1 and 2, though for very small h, say $h = 0.01$, clearly we would like some help from a computer or calculator.

Derivation of the diffusion equations. We follow the notation that has already been developed, except that values y_i will refer to quantity of heat rather than temperature (this will yield equations for temperature, since heat is a constant times temperature). What should happen at an interior node? The explanation requires one more experimentally observed law known as *Fourier's heat law*. It says that the flow of heat per unit length from one point to another is proportional to the rate of change in temperature with respect to distance and moves from higher temperature to lower. The constant of proportionality k is known as the *conductivity* of the material. In addition, we interpret the heat created at node x_i to be $hf(x_i)$, since f measures heat created per unit length. Count flow towards the right as positive. Thus, at node x_i the net flow per unit length from the left node and to the right node are given by

$$\text{Left flow} = k\frac{y_i - y_{i-1}}{h}$$

$$\text{right flow} = k\frac{y_i - y_{i+1}}{h}$$

Thus, in order to balance heat flowing through the ith node with heat created per unit length at this node, we should have

$$\text{Left flow} + \text{right flow} = k\frac{y_i - y_{i-1}}{h} + k\frac{y_i - y_{i+1}}{h} = hf(x_i)$$

In other words

$$k\frac{-y_{i-1} + 2y_i - y_{i+1}}{h^2} = f(x_i)$$

or

$$-y_{i-1} + 2y_i - y_{i+1} = \frac{h^2}{k}f(x_i) \qquad (3)$$

Input – Output models

We are going to set up a simple model of an economy consisting of three sectors that supply each other and consumers. Suppose the three sectors are (E)nergy, (M)aterials and (S)ervices and suppose that the demands of a sector are proportional to its output. This is reasonable: if, for example, the materials sector doubled its output, one would expect its needs for energy, material and services to likewise double. Now let x, y, z be the total output of the sectors E, M and S respectively. We require that the economy be *closed* in the sense that everything produced in the economy is consumed by the economy. Thus, the total output of the sector E should equal the amounts consumed by all the sectors and the consumers.

Example 4 Given the following table of demand constants of proportionality and consumer (D)emand (a fixed quantity) for the output of each service, express the closed property of the system as a system of equations.

		Consumed by			
		E	M	S	D
Produced by	E	0.2	0.3	0.1	2
	M	0.1	0.3	0.2	1
	S	0.4	0.2	0.1	3

Solution consider how we balance the total output and demands for energy. The total output is x units. The demands from the three sectors E, M and S are, according to the table data, $0.2x, 0.3y$ and $0.1z$, respectively. Further, consumers demand 2 units of energy. In equation form

$$x = 0.2x + 0.3y + 0.1z + 2$$

Likewise we can balance the input/output of the sectors M and S to arrive at a system of three equations in three unknowns.

$$x = 0.2x + 0.3y + 0.1z + 2$$
$$y = 0.1x + 0.3y + 0.1z + 1$$
$$z = 0.4x + 0.2y + 0.1z + 3$$

The questions that interest economists are whether or not this system has solutions, and if so, what they are.

Note In some of the text exercises you will find reference to "your computer system." This may be a calculator that is required for the course or a computer system for which you are given an account. This textbook does not depend on any particular system, but certain exercises require a computational device. The abbreviation "MAS" stands for a matrix algebra system like MATLAB or Octave. Also, the shorthand "CAS" stands for a computer algebra system like Maple, Mathematica or MathCad. A few of the projects are too large for most calculators and will require a CAS or MAS.

Exercise

1. Solve the following system algebraically.

(1) $\begin{cases} x + 2y = 1 \\ 3x - y = -4 \end{cases}$

(2) $\begin{cases} x - y + 2z = 6 \\ 2x - z = 3 \\ y + 2z = 0 \end{cases}$

(3) $\begin{cases} x - y = 1 \\ 2x - y = 3 \\ x + y = 3 \end{cases}$

2. Determine if the following systems of equations are linear or not. If so, put them in standard format

(1) $\begin{cases} x + 2 = y + z \\ 3x - y = 4 \end{cases}$

(2) $\begin{cases} xy + 2 = 1 \\ 2x - 6 = y \end{cases}$

(3) $\begin{cases} x + 2 = 1 \\ x + 3 = y^2 \end{cases}$

3. Express the following systems of equations in the notation of the definition of linear systems by specifying the numbers m, n, a_{ij} and b_i.

(1) $\begin{cases} x_1 - 2x_2 + x_3 = 2 \\ x_2 = 1 \\ -x_1 + x_3 = 1 \end{cases}$

(2) $\begin{cases} x_1 - 3x_2 = 1 \\ x_2 = 5 \end{cases}$

(3) $\begin{cases} x_1 - x_2 = 1 \\ 2x_1 - x_2 = 3 \\ x_2 + x_1 = 3 \end{cases}$

4. Write out the linear system that results from Example 3 if we take $n = 6$.

5. Suppose that in the input-output model of Example 4 we ignore the Materials sector input and output, so that there results a system of two equations in two unknowns x and z. Write out these equations and find a solution for them.

6. Here is an example of an economic system where everything

produced by the sectors of the system is consumed by those sectors. An administrative unit has four divisions serving the internal needs of the unit, labelled (A) ccounting, (M) aintenance, (S) uplies and (T)raining. Each unit produces the "commodity" its name suggests, and charges the other divisions for its services. The fraction of commodities consumed by each division is given by the following table, also called an "input – output matrix".

		Consumed by			
		A	M	S	T
	A	0.2	0.1	0.4	0.4
Consumed	M	0.3	0.4	0.2	0.1
	S	0.3	0.4	0.2	0.3
	T	0.2	0.1	0.2	0.2

7. A polynomial $y = a + bx + cx^2$ is required to interpolate a function $f(x)$ at $x = 1, 2, 3$ where $f(1) = 1, f(2) = 1$ and $f(3) = 2$. Express these three conditions as a linear system of three equations in the unknowns a, b, c.

Gaussian Elimination

A method for solving matrix equations of the form

$$Ax = b \qquad (1)$$

To perform Gaussian elimination starting with the system of equations

$$\begin{bmatrix} a_{11} & a_{12} & \cdots & a_{1k} \\ a_{21} & a_{22} & \cdots & a_{2k} \\ \vdots & \vdots & & \vdots \\ a_{k1} & a_{k2} & \cdots & a_{kk} \end{bmatrix} \begin{bmatrix} x_1 \\ x_2 \\ \vdots \\ x_k \end{bmatrix} = \begin{bmatrix} b_1 \\ b_2 \\ \vdots \\ b_k \end{bmatrix} \qquad (2)$$

compose the "augmented matrix equation"

$$\begin{bmatrix} a_{11} & a_{12} & \cdots & a_{1k} & b_1 \\ a_{21} & a_{22} & \cdots & a_{2k} & b_2 \\ \vdots & \vdots & & \vdots & \vdots \\ a_{k1} & a_{k2} & \cdots & a_{kk} & b_k \end{bmatrix} \begin{bmatrix} x_1 \\ x_2 \\ \vdots \\ x_k \end{bmatrix} \qquad (3)$$

Here, the column vector in the variables x is carried along for labeling the matrix rows. Now, perform elementary row and column operations to put the augmented matrix into the upper triangular form

$$\begin{bmatrix} a_{11}' & a_{12}' & \cdots & a_{1k}' & b_1' \\ 0 & a_{22}' & \cdots & a_{2k}' & b_2' \\ \vdots & \vdots & & \vdots & \vdots \\ 0 & 0 & \cdots & a_{kk}' & b_k' \end{bmatrix} \qquad (4)$$

Solve the equation of the kth row for x_k, then substitute back into the equation of the $(k-1)$st row to obtain a solution for x_{k-1}, etc., according to the formula

$$x_i = \frac{1}{a_{ii}'} \left(b_i' - \sum_{j=i+1}^{k} a_{ij}' x_j \right) \qquad (5)$$

For example, consider the matrix equation

$$\begin{bmatrix} 9 & 3 & 4 \\ 4 & 3 & 4 \\ 1 & 1 & 1 \end{bmatrix} \begin{bmatrix} x_1 \\ x_2 \\ x_3 \end{bmatrix} = \begin{bmatrix} 7 \\ 8 \\ 3 \end{bmatrix} \qquad (6)$$

In augmented form, this becomes

$$\begin{bmatrix} 9 & 3 & 4 & 7 \\ 4 & 3 & 4 & 8 \\ 1 & 1 & 1 & 3 \end{bmatrix} \begin{bmatrix} x_1 \\ x_2 \\ x_3 \end{bmatrix} \qquad (7)$$

Switching the first and third rows gives

$$\begin{bmatrix} 1 & 1 & 1 & 3 \\ 4 & 3 & 4 & 8 \\ 9 & 3 & 4 & 7 \end{bmatrix} \begin{bmatrix} x_1 \\ x_2 \\ x_3 \end{bmatrix} \qquad (8)$$

Subtracting 9 times the first row from the third row gives

$$\begin{bmatrix} 1 & 1 & 1 & | & 3 \\ 4 & 3 & 4 & | & 8 \\ 0 & -6 & -5 & | & -20 \end{bmatrix} \begin{bmatrix} x_1 \\ x_2 \\ x_3 \end{bmatrix} \tag{9}$$

Subtracting 4 times the first row from the second row gives

$$\begin{bmatrix} 1 & 1 & 1 & | & 3 \\ 0 & -1 & 0 & | & -4 \\ 0 & -6 & -5 & | & -20 \end{bmatrix} \begin{bmatrix} x_1 \\ x_2 \\ x_3 \end{bmatrix} \tag{10}$$

Finally, adding -6 times the second row to the third row gives

$$\begin{bmatrix} 1 & 1 & 1 & | & 3 \\ 0 & -1 & 0 & | & -4 \\ 0 & 0 & -5 & | & 4 \end{bmatrix} \begin{bmatrix} x_1 \\ x_2 \\ x_3 \end{bmatrix} \tag{11}$$

Restoring the transformed matrix equation gives

$$\begin{bmatrix} 1 & 1 & 1 \\ 0 & -1 & 0 \\ 0 & 0 & -5 \end{bmatrix} \begin{bmatrix} x_1 \\ x_2 \\ x_3 \end{bmatrix} = \begin{bmatrix} 3 \\ -4 \\ 4 \end{bmatrix} \tag{12}$$

which can be solved immediately to give $x_3 = \dfrac{-4}{5}$, back-substituting to obtain $x_2 = 4$ (which actually follows trivially in this example), and then again back-substituting to find $x_1 = \dfrac{-1}{5}$.

Words and Expressions

linear algebra 线性代数
least *adj*. 最小的,最少的 *adv*. 最小,最少
fit *vt*. 安装,装备;使适应
fitting *n*. 拟合
least square fitting 最小二乘方拟合
solution *n*. 解,解法
differential *n*. 微分,微分的
differential equations 微分方程
matrix *n*. *pl*. *matrices or matrixes* 矩阵,真值表

determinant $n.$ 行列式
matrix equation 矩阵方程
matrix inverse 矩阵的逆
Gaussian 高斯
elimination $n.$ 消元法,消去
Gaussian Elimination 高斯消元法
robust $n.$ 鲁棒 $adj.$ 稳健
ring $n.$ 环
axiom $n.$ 公理,公设
Linear Systems of Equations 线性方程组
square matrix 方阵
sophisticated $adj.$ 复杂的,完善的,成熟的
dimension $n.$ 维,维数;量纲
three dimensions 三维
identical $adj.$ 恒等的,恒同的
distinct $adj.$ 分开的,不同的
parallel $adj.$ 平行的,并行的 $n.$ 平行线
intersect $vt.$ 交,相交
subtract $vt.$ 减去
constant $n.$ 常项,常数
constant term 常数项
coefficient $n.$ 系数
index $n.$ 指标;指数;下标;索引 $vt.$ 标引
row $n.$ 行
ith row 第 i 行
homogeneous $adj.$ 齐次的,齐的
proportional $adj.$ 成比例的,相称的
range $n.$ 值域,范围,变程
diffusion $n.$ 扩散
per unit length 每个长度单位

respectively *adv.* 各自地,个别地
matrix equation 矩阵方程
upper triangular form 上三角形
trivially *adv.* 平凡地

3.3 MATRIX ALGEBRA

Our notational style of writing a matrix in the from $A = [a_{ij}]$ hints that a matrix could be treated like a single number. What if we could manipulate equations with matrix and vector quantities in the same way that we do scalar equations? We shall see that is a useful idea. Matrix arithmetic gives us new powers for formulating and solving practical problems. In the following we will develop the arithmetic of matrices and vectors. It will help us to find effective methods for solving linear and nonlinear systems, solve problems of graph theory.

Matrix Addition and Scalar Multiplication

To begin our discussion of arithmetic we consider the matter of equality of matrices. Suppose that A and B represent two matrices. When do we declare them to be equal? The answer is, of course, if they represent the same matrix. Thus we expect that all the usual laws of equalities will hold (e.g., equals may be substituted for equals) and in fact, they do. There are times, however, when we need to prove that two symbolic matrices must be equal. For this purpose, we need something a little more precise. So we have the following definition, which includes vectors as a special case of matrices.

Definition 1 Two matrices $A = [a_{ij}]$ and $B = [b_{ij}]$ are said to be *equal* if these matrices have the same size, and for each index pair (i, j), $a_{ij} = b_{ij}$, that is, corresponding entries of A and B are equal.

Example 1 Which of the following matrices are equal, if any?

(1) $\begin{bmatrix} 0 \\ 0 \end{bmatrix}$ (2) $[0,0]$ (3) $\begin{bmatrix} 0 & 1 \\ 0 & 2 \end{bmatrix}$ (3) $\begin{bmatrix} 0 & 1 \\ 1-1 & 1+1 \end{bmatrix}$

Solution The answer is that only (3) and (4) have any chance of being equal, since they are the only matrices in the list with the same size (2×2). As a matter of fact, an entry check verifies that they really are equal.

Matrix Inverses

We have seen that if we could make sense of "$\frac{1}{A}$", then we could write the solution to the linear system $Ax = b$ as simply $x = \left(\frac{1}{A}\right) b$. We are going to tackle this problem now. First, we need a definition of the object that we are trying to uncover. Notice that "inveres" could only work on one side. For example

$$[1 \quad 2]\begin{bmatrix} -1 \\ 1 \end{bmatrix} = [1] = [2 \quad 3]\begin{bmatrix} -1 \\ 1 \end{bmatrix}$$

which suggests that both $[1 \quad 2]$ and $[2 \quad 3]$ should qualify as left inverses of the matrix $\begin{bmatrix} -1 \\ 1 \end{bmatrix}$, since multiplication on the left by them results in a 1×1 identity matrix. As a matter of fact right and left inverses are studied and do have applications. But they have some unusual properties such as non-uniqueness. We are going to focus on a type of inverse that is closer to the familiar inverses in fields of numbers, namely, *two-sided* inverses. These only make sense for square matrices, so the non-square example above is ruled out.

Definition Let A be a square matrix. Then a (*two-sided*) *inverse* for A is a square matrix B of the same size as A such that $AB = I = BA$. If such a B exists, then the matrix A is said to be *invertible*.

Of course, any non-square matrix is non-invertible. Square matrices are classified as either "singular", i. e., non-invertible, or "nonsingular", i.e., invertible. Since we will mostly be concerned with two-sided inverses, the unqualified term "inverse" will be understood to mean a "two-sided inverse." Notice that this definition is actually symmetric in A and B. In other words, if B is an inverse for A, then A is an inverse for B.

Example Show that $B = \begin{bmatrix} 1 & 1 \\ 1 & 2 \end{bmatrix}$ is an inverse for the matrix $A = \begin{bmatrix} 2 & -1 \\ -1 & 1 \end{bmatrix}$.

Solution All we have to do is to check the definition. But remember that there are two multiplications to confirm. (We'll show later that this isn't necessary, but right now we are working strictly from the definition.) We have

$$AB = \begin{bmatrix} 2 & -1 \\ -1 & 1 \end{bmatrix}\begin{bmatrix} 1 & 1 \\ 1 & 2 \end{bmatrix} = \begin{bmatrix} 2\cdot 1 - 1\cdot 1 & 2\cdot 1 - 1\cdot 2 \\ -1\cdot 1 + 1\cdot 1 & -1\cdot 1 + 1\cdot 2 \end{bmatrix} = \begin{bmatrix} 1 & 0 \\ 0 & 1 \end{bmatrix} = I$$

Words and Expressions

matrix algebra 矩阵代数
matrix addition 矩阵加法
multiplication n. 乘法
scalar multiplication 数量乘法
equality n. 等式,相等
equal vt. 等于,相等 $adj.$ 相等的
symbolic $adj.$ 符号的
invertible $n.$ 可逆的
non-square matrix 非方阵
non-invertible 非可逆的

3.4 EIGENVALUES AND EIGENVECTORS

Have you ever heard the words eigenvalue and eigenvector? They are derived from the German word "eigen" which means "propers" or "characteristic". An eigenvalue of a square matrix is a scalar that is usually represented by the Greek letter λ (pronounced lambda). As you might suspect, an eigenvector is a vector. Moreover, we require that an eigenvector be a nonzero vector, in other words, an eignvector can not be the zero vector. We will denote an eigenvector by the small letter x. All eigenvalues and eigenvectors satisfy the equation $Ax = \lambda x$ for a given square matrix A.

Remark 1   Remember that, in general, the word scalar is not restricted to real numbers. We are only using real numbers as scalars in this book, but eigenvalues are often complex numbers.

Definition 1   Consider the square matrix A. We say that λ is an eigenvalue of A if there exists a non-zero vector x such that $Ax = \lambda x$. In this case, x is called an eigenvector (corresponding to λ), and the pair (λ, x) is called an eigenpair for A.

Let's look at an example of an eigenvalue and eigenvector. If you were asked if $x = \begin{bmatrix} 1 \\ -2 \end{bmatrix}$ is an eigenvector corresponding to the eigenvalue $\lambda = 0$ for $A = \begin{bmatrix} 6 & 3 \\ -2 & -1 \end{bmatrix}$, you could find out by substituting x, λ and A into the equation $Ax = \lambda x$, we have

$$\begin{bmatrix} 6 & 3 \\ -2 & -1 \end{bmatrix} \begin{bmatrix} 1 \\ -2 \end{bmatrix} = 0 \begin{bmatrix} 1 \\ -2 \end{bmatrix}$$

$$\begin{bmatrix} 0 \\ 0 \end{bmatrix} = \begin{bmatrix} 0 \\ 0 \end{bmatrix}$$

Therefore, λ and x are an eigenvalue and an eigenvector, respectively, for A.

Now that you have seen an eigenvalue and an eigenvector, let's talk a little more about them. Why did we require that an eigenvector not be zero? If the eigenvector was zero, the equation $Ax = \lambda x$ would yield $0 = 0$. Since this equation is always true, it is not an interesting case. Therefore, we define an eigenvector to be non-zero vector that satisfies $Ax = \lambda x$.

What should we do if we want to find all the eigenpairs? We know that $Ax = \lambda x$ for all eigenpairs. We can transform this equation into a form that will help us. When we learned about the identity matrix, we learned that $Ix = x$ for any x. Therefore, $Ax = \lambda Ix$. We can use algebra steps from here

$$Ax = \lambda Ix$$
$$Ax - \lambda Ix = 0$$
$$(A - \lambda I)x = 0$$

We know that $x = 0$ would solve this equation, but we defined an eigenvector to be non zero, so if there is an eigenvector solution to the equation $(A - \lambda I)x = 0$ then there must be more than one solution to the equation. We learned that the system has a unique solution if $\det(A - \lambda I) \neq 0$. Therefore, we know that if there is a non zero solution to $(A - \lambda I)x = 0$ then $\det(A - \lambda I) = 0$. The equation $\det(A - \lambda I) = 0$ even has a name. It is called the characteristic equation. We can solve the characteristic equation to find all the eigenvalues of certain matrices. There will be as many eigenvalues as there are rows in the matrix (or columns since the matrix must be square), but some of the eigenvalues might be identical to each other.

Let's find both of the eigenvalues of the matrix $A = \begin{bmatrix} 3 & 6 \\ 1 & 4 \end{bmatrix}$

$$A - \lambda I = \begin{bmatrix} 3 - \lambda & 6 \\ 1 & 4 - \lambda \end{bmatrix}$$

$$\det(A - \lambda I) = (3 - \lambda)(4 - \lambda) - 6 =$$
$$\lambda^2 - 7\lambda + 6 = (\lambda - 6)(\lambda - 1)$$

Therefore, $\lambda = 6$ or $\lambda = 1$. We now know our eigenvalues. Remember that all eigenvalues are paired with an eigenvector.

Therefore, we can substitute our eigenvalues, one at a time, into the formula $(A - \lambda I)x = 0$ and solve to find a corresponding eigenvector.

Let's find an eigenvector corresponding to $\lambda = 6$

$$(A - \lambda I)x = 0$$

$$\left(\begin{bmatrix} 3 & 6 \\ 1 & 4 \end{bmatrix} - \begin{bmatrix} 6 & 0 \\ 0 & 6 \end{bmatrix}\right)x = \begin{bmatrix} 0 \\ 0 \end{bmatrix}$$

$$\begin{bmatrix} -3 & 6 \\ 1 & -2 \end{bmatrix} x = \begin{bmatrix} 0 \\ 0 \end{bmatrix}$$

$$\left[\begin{array}{cc|c} -3 & 6 & 0 \\ 1 & -2 & 0 \end{array}\right] \text{Augmented Matrix}$$

$$\left[\begin{array}{cc|c} 1 & -2 & 0 \\ 1 & -2 & 0 \end{array}\right] r_1 \div (-3)$$

$$\left[\begin{array}{cc|c} 1 & -2 & 0 \\ 0 & 0 & 0 \end{array}\right] -1 \times r_1 \times r_2$$

Notice that this system is underdetermined. Therefore, there are an infinite number of solutions. So, any vector that solves the equation $x_1 - 2x_2 = 0$ is an eigenvector corresponding to $\lambda = 6$ when $A = \begin{bmatrix} 3 & 6 \\ 1 & 4 \end{bmatrix}$. To have a consistent method for finding an eigenvector, let's choose the solution in which $x_2 = 1$. We can use back-substitution to find that $x_1 - 2 \times (1) = 0$ which implies that $x_1 = 2$. This tells us that $\begin{bmatrix} 2 \\ 1 \end{bmatrix}$ is an eigenvector corresponding to $\lambda = 6$ when $A = \begin{bmatrix} 3 & 6 \\ 1 & 4 \end{bmatrix}$. This is the same solution that we found when we used the power method to find the dominant eigenpair.

Let's find an eigenvector corresponding to $\lambda = 1$

$$(A - \lambda I)x = 0$$

$$\left(\begin{bmatrix} 3 & 6 \\ 1 & 4 \end{bmatrix} - \begin{bmatrix} 1 & 0 \\ 0 & 1 \end{bmatrix}\right)x = \begin{bmatrix} 0 \\ 0 \end{bmatrix}$$

$$\begin{bmatrix} 2 & 6 \\ 1 & 3 \end{bmatrix} x = \begin{bmatrix} 0 \\ 0 \end{bmatrix}$$

$$\left[\begin{array}{cc|c} 2 & 6 & 0 \\ 1 & 3 & 0 \end{array}\right] \text{ Augmented Matrix}$$

$$\left[\begin{array}{cc|c} 1 & 3 & 0 \\ 1 & 3 & 0 \end{array}\right] r_1 \div 2$$

$$\left[\begin{array}{cc|c} 1 & 3 & 0 \\ 0 & 0 & 0 \end{array}\right] - 1 \times r_1 \times r_2$$

Notice that this system is underdetermined. This will always be true when we are finding an eigenvector using this method. So, any vector that solves the equation $x_1 + 3x_2 = 0$ is an eigenvector corresponding to the eigenvalue $\lambda = 1$ when $A = \begin{bmatrix} 3 & 6 \\ 1 & 4 \end{bmatrix}$. Again, let's choose the eigenvector in which the last element of x is 1, Therefore, $x_2 = 1$ and $x_1 + 3 \times (1) = 0$, so $x_1 = -3$. This tells us that $\begin{bmatrix} -3 \\ 1 \end{bmatrix}$ is an eigenvector corresponding to $\lambda = 1$ when $A = \begin{bmatrix} 3 & 6 \\ 1 & 4 \end{bmatrix}$. Using the characteristic equation and Gaussian elimination, we are able to find all the eigenvalues to the matrix and corresponding eigenvectors.

Let's find the eigenpairs for the matrix $A = \begin{bmatrix} 2 & 6 \\ 2 & -2 \end{bmatrix}$ for which the power method fails

$$A - \lambda I = \begin{bmatrix} 2 - \lambda & 6 \\ 2 & -2 - \lambda \end{bmatrix}$$

$$\det(A - \lambda I) = (2 - \lambda)(-2 - \lambda) - 12$$

$$\lambda^2 - 16 = (\lambda - 4)(\lambda + 4)$$

Therefore, $\lambda = 4$ or $\lambda = -4$. The power method does not work because $|4| = |-4|$. In other words, there is not a unique dominant eigenvalue.

Let's find an eigenvector corresponding to $\lambda = 4$

$$(A - \lambda I)x = 0$$

$$\left(\begin{bmatrix} 2 & 6 \\ 2 & -2 \end{bmatrix} - \begin{bmatrix} 4 & 0 \\ 0 & 4 \end{bmatrix}\right)x = \begin{bmatrix} 0 \\ 0 \end{bmatrix}$$

$$\begin{bmatrix} -2 & 6 \\ 2 & -6 \end{bmatrix} x = \begin{bmatrix} 0 \\ 0 \end{bmatrix}$$

$$\left[\begin{array}{cc|c} -2 & 6 & 0 \\ 2 & -6 & 0 \end{array}\right] \text{Augmented Matrix}$$

$$\left[\begin{array}{cc|c} 1 & -3 & 0 \\ 2 & -6 & 0 \end{array}\right] r_1 \div (-2)$$

$$\left[\begin{array}{cc|c} 1 & -3 & 0 \\ 0 & 0 & 0 \end{array}\right] -2r_1 \times r_2$$

Since the system is underdetermined, we have an infinite number of solutions. Let's choose the solution in which $x_2 = 1$. We can use back-substitution to find that $x_1 - 3 \times (1) = 0$ which implies that $x_1 = 3$. This tells us that $\begin{bmatrix} 3 \\ 1 \end{bmatrix}$ is an eigenvector corresponding to $\lambda = 4$ when $A = \begin{bmatrix} 2 & 6 \\ 2 & -2 \end{bmatrix}$.

Let's find an eigenvector corresponding to $\lambda = -4$

$$(A - \lambda I)x = 0$$

$$\left(\begin{bmatrix} 2 & 6 \\ 2 & -2 \end{bmatrix} - \begin{bmatrix} -4 & 0 \\ 0 & -4 \end{bmatrix}\right)x = \begin{bmatrix} 0 \\ 0 \end{bmatrix}$$

$$\begin{bmatrix} 6 & 6 \\ 2 & 2 \end{bmatrix} x = \begin{bmatrix} 0 \\ 0 \end{bmatrix}$$

$$\left[\begin{array}{cc|c} 6 & 6 & 0 \\ 2 & 2 & 0 \end{array}\right] \text{Augmented Matrix}$$

$$\begin{bmatrix} 1 & 1 & | & 0 \\ 2 & 2 & | & 0 \end{bmatrix} r_1 \div 6$$

$$\begin{bmatrix} 1 & 1 & | & 0 \\ 0 & 0 & | & 0 \end{bmatrix} -2r_1 \times r_2$$

Again, let's choose the eignvector in which the last element of x is 1. Therefore, $x_2 = 1$ and $x_1 + 1 \times (1) = 0$, so $x_1 = -1$. This tells us that $\begin{bmatrix} -1 \\ 1 \end{bmatrix}$ is an eignvector corresponding to $\lambda = -4$ when $A = \begin{bmatrix} 2 & 6 \\ 2 & -2 \end{bmatrix}$. Using the characteristic equation and Gaussian elimination, we are able to find both of the eigenvalues to the matrix and corresponding eigenvectors.

We can find eigenpairs for larger systems using this method, but the characteristic equation gets impossible to solve directly when the system gets too large. We could use approximations that get close to solving the characteristic equation, but there are better ways to find eigenpairs that you will study in the future. However, these two methods give you an idea of how to find eigenpairs.

Another matrix for which the power method will not work is the matrix $A = \begin{bmatrix} 5 & 0 \\ 0 & 5 \end{bmatrix}$ because the eigenvalues are both the real number 5. The method that we showed you earlier will yield the eigenvector $\begin{bmatrix} 0 \\ 1 \end{bmatrix}$ to correspond to the eigenvalue $\lambda = 5$. Other methods will reveal, and you can check, that $\begin{bmatrix} 0 \\ 1 \end{bmatrix}$ is also an eigenvector of A corresponding to $\lambda = 5$. Notice that these two eigenvectors are not multiples of one another. If the same eigenvalue is repeated p times for a particular matrix, then there can be as many as p different eigenvectors that are not multiples of each other that correspond to that eigenvalue.

We said that eigenvalues are often complex numbers. However, if the matrix A is symmetric, then the eigenvalues will always be real numbers. As you can see from the problems that we worked, eigenvalues can also be real when the matrix is not symmetric, but keep in mind that they are not guaranteed to be real.

Did you know that the determinant of a matrix is related to the eigenvalues of the matrix? The product of the eigenvalues of a square matrix is equal to the determinant of that matrix. Let's look at the two matrices that we have been working with, for $\begin{bmatrix} 3 & 6 \\ 1 & 4 \end{bmatrix}$

$$\text{Product of eigenvalues} = \det(A)$$
$$6 \times 1 = 12 - 6$$
$$6 = 6$$

For $\begin{bmatrix} 2 & 6 \\ 2 & -2 \end{bmatrix}$

$$\text{Product of eigenvalues} = \det(A)$$
$$6 \times 1 = 12 - 6$$
$$6 = 6$$

You can use this as a check to see that you have the correct eigenvalues and determinant for the matrix A.

Now that we know how to find eigenpairs, we might want to know what uses they have. The interesting uses come from larger systems, so we will just discuss them rather than solve them.

Words and Expressions

eigenvalue $n.$ 特征值

eigenvector $n.$ 特征向量

derive $vt.$ 获得,得来;起源,由来 $vi.$ 起源,发生

derived from 起源,发生

characteristic $adj.$ 特征的,示性的

remark $n.$ 注意,意见,附注(常用复数)

restricted *adj.* 限制的,约束的
correspond *vi.* 对应
corresponding *adj.* 同位的,对应的(与 *to* 连用)
eigenpair *n.* 特征对,本征对
substitute *vt.* 代换,代入 *n.* 代替者,代理人;代替物,替代品
transform *vt.* 使变形
identity *n.* 恒等,恒等式,单位元
identity matrix 单位矩阵
step *n.* 阶段,级,步骤,步长
algebra steps 代数步骤
unique *adj.* 惟一的
unique solution 惟一解
non-zero solution 非零解
characteristic equation 特征方程
identical *adj.* 恒等的,恒同的
augment *vt. or vi.* 增大,增加
augmented matrix 增广矩阵
underdetermined *adj.* 欠定,亚定
consistent *adj.* 相容的,一致的,无矛盾的
consistent method 一致的方法
dominant *adj.* 支配的
dominant eigenpair 主特征对,主本征对
multiples *n.* 倍数,倍式;多重
symmetric *adj.* 对称的
guarantee *n.* 担保,保证;担保人,保证人;保证书 *vt.* 保证,担保
product of eigenvalues 特征值的积

4

Probability

4.1 INTRODUCTION TO PROBABILITY

Problem A spinner has 4 equal sectors marked a, b, c and d. What are the chances of landing on the sector b after spinning the spinner?

Solution The chances of landing on the sector b are 1 in 4, or one fourth.

This problem asked us to find the probability that the spinner will land on the sector b. Let's look at some definitions and examples from the problem above.

Definition	Example
An experiment is a situation involving chance or probability that leads to results called outcomes.	The experiment is spinning the spinner.
An outcome is the result of a single trial of an experiment.	The possible outcome are landing on the sector a, b, c or d.

Definition	Example
An event is one or more outcomes of an experiment.	The event being measured is landing on the sector b.
Probability is the measure of how likely an event is.	The probability of landing on the sector b is one fourth.

In order to measure probabilities, mathematicians have devised the following formula for finding the probability of an event.

Probability Of An Event

$$P(A) = \frac{\text{The Number Of Ways Event A Can Occur}}{\text{The Total Number Of Possible Outcomes}}$$

The Probability of event A is the number of ways event A can occur divided by the total number of possible outcomes. Let's take a look at a slight modification of the problem from the above.

Experiment 1 A spinner has 4 equal sectors marked a, b, c and d. After spinning the spinner, what is the probability of landing on each sector?

Outcomes The possible outcomes of this experiment are a, b, c and d.

Probabilities

$$P(a) = \frac{\text{number of ways to land on } a}{\text{total number of sectors}} = \frac{1}{4}$$

$$P(b) = \frac{\text{number of ways to land on } b}{\text{total number of sectors}} = \frac{1}{4}$$

$$P(c) = \frac{\text{number of ways to land on } c}{\text{total number of sectors}} = \frac{1}{4}$$

$$P(d) = \frac{\text{number of ways to land on } d}{\text{total number of sectors}} = \frac{1}{4}$$

Experiment 2 A single 6-sided die is rolled. What is the probability of each outcome? What is the probability of rolling an even number (2, 4 or 6)? an odd number (1, 3 or 5)?

Outcomes The possible outcomes of this experiment are 1, 2, 3, 4, 5 and 6.

$$P(1) = \frac{\text{number of ways to roll a 1}}{\text{total number of sides}} = \frac{1}{6}$$

$$P(2) = \frac{\text{number of ways to roll a 2}}{\text{total number of sides}} = \frac{1}{6}$$

$$P(3) = \frac{\text{number of ways to roll a 3}}{\text{total number of sides}} = \frac{1}{6}$$

$$P(4) = \frac{\text{number of ways to roll a 4}}{\text{total number of sides}} = \frac{1}{6}$$

$$P(5) = \frac{\text{number of ways to roll a 5}}{\text{total number of sides}} = \frac{1}{6}$$

$$P(6) = \frac{\text{number of ways to roll a 6}}{\text{total number of sides}} = \frac{1}{6}$$

$$P(even) = \frac{\text{ways to roll an even number}}{\text{total number of sides}} = \frac{3}{6} = \frac{1}{2}$$

$$P(odd) = \frac{\text{ways to roll an odd number}}{\text{total number of sides}} = \frac{3}{6} = \frac{1}{2}$$

This experiment illustrates the diffrerence between an outcome and an event. A single outcome of this experiment is rolling a 1, or a 2, or a 3, or a 4, or a 5, or a 6. Rolling an even number (2, 4 or 6) is an event, and rolling an odd number (1, 3 or 5) is also an event.

In Experiment 1 the probability of each outcome is always the same. The probability of landing on each sector of the spinner is always one fourth. In Experiment 2, the probability of rolling each number on the die is always one sixth. In each of these experiments the outcomes are equally likely to occure. Let's look at an experiment in which the outcomes are not equally likely.

Experiment 3   A glass jar contain 6 red, 5 green, 8 blue, and 3 yellow marbles. If a single marble is chosen at random from the jar, what is the probability that it is red? green? blue? yellow?

Outcomes The possible outcomes of this experiment are red, green, blue, and yellow.

Probabilities

$$P(red) = \frac{\text{number of ways to choose red}}{\text{total number of marbles}} = \frac{6}{22} = \frac{3}{11}$$

$$P(green) = \frac{\text{number of ways to choose green}}{\text{total number of marbles}} = \frac{5}{22}$$

$$P(blue) = \frac{\text{number of ways to choose blue}}{\text{total number of marbles}} = \frac{8}{22} = \frac{4}{11}$$

$$P(yellow) = \frac{\text{number of ways to choose yellow}}{\text{total number of marbles}} = \frac{3}{22}$$

The outcomes in this experiment are not equally likely to occur. You are much more likely to choose a blue marble than any other color marble. You are least likely to choose a yellow marble.

Experiment 4 Choose a number at random from 1 to 5. What is the probability of each outcome? What is the probability that the number chosen is even? odd?

Outcomes The possible outcomes of this experiment are 1, 2, 3, 4, and 5.

Probabilities

$$P(1) = \frac{\text{number of ways to choose 1}}{\text{total number of numbers}} = \frac{1}{5}$$

$$P(2) = \frac{\text{number of ways to choose 2}}{\text{total number of numbers}} = \frac{1}{5}$$

$$P(3) = \frac{\text{number of ways to choose 3}}{\text{total number of numbers}} = \frac{1}{5}$$

$$P(4) = \frac{\text{number of ways to choose 4}}{\text{total number of numbers}} = \frac{1}{5}$$

$$P(5) = \frac{\text{number of ways to choose 5}}{\text{total number of numbers}} = \frac{1}{5}$$

$$P(even) = \frac{\text{number of ways to choose an even number}}{\text{total number of numbers}} = \frac{2}{5}$$

$$P(odd) = \frac{\text{number of ways to choose an odd number}}{\text{total number of numbers}} = \frac{3}{5}$$

The outcomes 1, 2, 3, 4 and 5 are equally likely to occur as a result of this experiment. However, the event even and odd are not equally likely to occur, since there are 3 odd numbers and only 2 even number from 1 to 5.

Summary

The probability of an event is the measure of the chance that the event will occur as a result of an experiment. The probability of an event A is the number of ways event A can occur divided by the total number of possible outcomes. The probability of an event A, symbolized by $P(A)$, is a number between 0 and 1, inclusive, that measures the likelihood of an event in the following way:

· $P(A) > P(B)$ then event A is more likely to occur than event B.
· $P(A) = P(B)$ then events A and B are equally likely to occur.

Probability is the branch of mathematics which studies the possible outcomes of given events together with their relative likelihoods and distributions. In common usage, the word "probability" is used to mean the chance that a particular event (or set of events) will occur expressed on a linear scale from 0 (impossibility) to 1 (certainty), also expressed as a percentage between 0 and 100%. The analysis of events governed by probability is called statistics.

There are several competing interpretations of the actual "meaning" of probabilities. Frequentists view probability simply as a measure of the frequency of outcomes (the more conventional interpretation), while Bayesians treat probability more subjectively as a statistical procedure which endeavors to estimate parameters of an underlying distribution based on the observed distribution.

A properly normalized function which assigns a probability "density" to each possible outcome within some interval is called a probability function (or probability distribution function), and its cumulative value (integral for a continuous distribution or sum for a discrete distribution) is called a distribution function (or cumulative distribution function).

A variate is defined as the set of all random variables that obey a given probabilistic law. It is common practice to denote a variate with a capital letter (most commonly X). The set of all values that X can take is then called the range, denoted R_X. Specific elements in the range of X are called quantiles and denoted x, and the probability that a variate X assumes the element x is denoted $P(X = x)$.

Probabilities are defined to obey certain assumptions, called the probability axioms. Let a sample space contain the union ($\bigcup$) of all possible events E_i, so

$$S \equiv \left(\sum_{i=1}^{N} E_i \right) \quad (1)$$

and let E and F denote subsets of S. Further, let $F' = \text{not} - F$ be the complement of F, so that

$$F \bigcup F' = S \quad (2)$$

Then the set E can be written as

$$E = E \bigcap S = E \bigcap (F \bigcup F') = (E \bigcap F) \bigcup (E \bigcap F') \quad (3)$$

where $\bigcap$ denotes the intersection. Then

$$\begin{aligned}
P(E) &= P(E \bigcap F) + P(E \bigcap F') - P[(E \bigcap F) \bigcap (E \bigcap F')] = \\
&\quad P(E \bigcap F) + P(E \bigcap F') - P[(F \bigcap F') \bigcap (E \bigcap E)] = \\
&\quad P(E \bigcap F) + P(E \bigcap F') - P[\phi \bigcap (E)] = \\
&\quad P(E \bigcap F) + P(E \bigcap F') - P[\phi] = \\
&\quad P(E \bigcap F) + P(E \bigcap F') \quad (4)
\end{aligned}$$

where ϕ is the empty set.

Let $P(E|F)$ denote the conditional probability of E given that F has already occurred, then

$$P(E) = P(E|F)P(F) + P(E|F')P(F') = \quad (5)$$

$$P(E|F)P(F) + P(E|F')[1 - P(F)] \quad (6)$$

$$P(A \bigcap B) = P(A)P(B|A) = \quad (7)$$

$$P(B)P(A|B) \quad (8)$$

$$P(A' \cap B) = P(A')P(B|A') \qquad (9)$$

$$P(E|F) = \frac{P(E \cap F)}{P(F)} \qquad (10)$$

The relationship

$$P(A \cap B) = P(A)P(B) \qquad (11)$$

holds if A and B are independent events. A very important result states that

$$P(E \cup F) = P(E) + P(F) - P(E \cap F) \qquad (12)$$

which can be generalized to

$$P(\bigcup_{i=1}^{n} A_i) = \sum^{i} P(A_i) - \sum_{ij}{}' P(A_i \cup A_j) + \sum_{ijk}{}'' P(A_i \cap A_j \cap A_k) - \cdots + (-1)^{n-1} P(\bigcap_{i=1}^{n} A_i) \qquad (13)$$

Exercises

Directions: Read each question below. Select your answer.

1. Which of the following is an experiment?
    (1) Tossing a coin   (2) Rolling a single die
    (3) Choosing a marble from a jar (4) All of the above

2. Which of the following is an outcome?
    (1) Rolling a pair of dice   (2) landing on red
    (3) Choosing 2 marbles from a jar (4) None of the above

3. Which of the following experiments does NOT have equally likely outcomes?
    (1) Choose a number at random from 1 to 7
    (2) Toss a coin
    (3) Choose a letter at random from the word SCHOOL
    (4) None of the above

4. What is the probability of choosing a vowel from the alphabet?
    (1) $\frac{21}{26}$ (2) $\frac{5}{26}$

(3) $\frac{1}{26}$ (4) None of the above

5. A number from 1 to 11 is chosen at random. What is the probability of choosing an odd number?

(1) $\frac{1}{11}$ (2) $\frac{5}{11}$

(3) $\frac{6}{11}$ (4) None of the above

Words and Expressions

probability *n.* 概率
spin *vt.* 自转,旋转
spinner *n.* 旋转物
chance *n.* 机会
land *n.* 陆地,国土 *vt.* 使降落,使着陆 *vi.* 登岸,着陆
land on 落在(一个较小的)物体上
experiment *n.* 实验,试验
outcome *n.* 结果
trial *n.* 试,试验
event *n.* 事件
formula *n.* 公式
total number 总数
possible *adj.* 可能的
possible outcome 可能的结果
total number of possible outcome 所有可能的结果
die *n.* 骰子 *pl.* dice ;小立方体;印膜
even *adj.* 偶的,偶数的;整数的
even number 偶数
odd *adj.* 奇数的
odd number 奇数
difference *n.* 差,差分

random *adj.* 随机的
equally *adv.* 相等地,同样地
likely *n.* 可能的
equally likely 等可能的
symbolize *vt.* 以符号表示
likelihood *n.* 似然
distribution *n.* 分布,分配;
statistics *n.* 统计,统计学
parameter *n.* 参数,参变量
estimate *vt.* 估计 *n.* 估计量
probability function（**or probability distribution function**）概率函数
cumulative *adj.* 累积的
continuous distribution（**or cumulative distribution function**）连续分布
random variables 随机变量

5

Selected Problems

5.1 THE REAL NUMBER SYSTEMS

Ordered Sets

Definition Let S be a set. An *order* on S is a relation, denoted by $<$, with the following two properties:

(i) If $x \in S$ and $y \in S$ then one and only one of the statements
$$x < y, \quad x = y, \quad y < x$$
is true.

(ii) If $x, y, z \in S$, if $x < y$ and $y < z$ then $x < z$.

The statement "$x < y$" may be read as "x is less than y" or "x is smaller than y" or "x precedes y".

It is often convenient to write $y > x$ in place of $x < y$.

The notation $x \leq y$ indicates that $x < y$ or $x = y$, without specifying which of these two is to hold. In other words, $x \leq y$ is the negation of $x > y$.

Fields

Definition   A *field* is a set F with two operations, called *addition* and *multiplication*, which satisfy the following so-called "field axioms" (A), (M), and (D):

(A) Axioms for addition

(A_1) If $x \in F$ and $y \in F$, then their sum $x + y$ is in F.

(A_2) Addition is commutative: $x + y = y + x$ for all $x, y \in F$.

(A_3) Addition is associative: $(x + y) + z = x + (y + z)$ for all $x, y, z \in F$.

(A_4) F contains an element 0 such that $0 + x = x$ for every $x \in F$.

(A_5) To every $x \in F$ corresponds an element $-x \in F$ such that
$$x + (-x) = 0.$$

(M) Axioms for multiplication

(M_1) If $x \in F$ and $y \in F$, then their product xy is in F.

(M_2) Multiplication is commutative: $xy = yx$ for all $x, y \in F$.

(M_3) Multiplication is associative: $(xy)z = x(yz)$ for all $x, y, z \in F$.

(M_4) F contains an element $1 \neq 0$ such that $1x = x$ for every $x \in F$.

(M_5) If $x \in F$ and $x \neq 0$ then there exists an element $\frac{1}{x} \in F$ such that $x \cdot \frac{1}{x} = 1$.

(D) The distributive law

$$x(y + z) = xy + xz$$

holds for all $x, y, z \in F$.

The Real Field

We now state the existence theorem which is the core of this chapter.

Theorem There exists an ordered field R which has the least-upper-bound property.

Moreover, R contains Q as subfield.

The second statement means that $Q \subset R$ and the operation of addition and multiplication in R, when applied to members of Q, coincide with the usual operations on rational numbers; also, the positive rational numbers are positive element of R.

The members of R are called real numbers.

The proof of Theorem as rather long and a bit tedious. The proof actually constructs R from Q.

The next theorem could be extracted from this construction with very little extra effort. However, we prefer to derive it from Theorem since this provides a good illustration of what one can do with the least-upper-bound property.

Theorem

(a) If $x \in R$, $y \in R$, and $x > 0$, then there is a positive integer n such that
$$nx > y.$$

(b) If $x \in R$, $y \in R$, and $x < y$, then there exists a $p \in Q$ such that $x < p < y$.

Words and Expressions

order n. 序,次;阶,级 vt. 定货
set n. 集,集合
ordered set 有序集
relation n. 关系
precede vt. 在前,在先,优于 vi. 在前,在先
indicate vt. 指示,指出;显示;表示
specify vt. or vi. 指定,载明
negation n. 否定,否定词

field *n*. 域
operation *n*. 运算
addition *n*. 加法
multiplication *n*. 乘法
commutative *adj*. 交换的
associative *adj*. 结合的
distributive *adj*. 分配的
distributive law 分配率
core *n*. 核心,柱心
least upper bound 上确界
property *n*. 性质
subfield *n*. 子域
coincide *vi*. 一致,符合;空间上相合;时间上相合

5.2 SELECTED PROBLEMS FROM THE INFINITE SERIES

Mathematicians have been intrigued by Infinite Series ever since antiquity. The question of how an infinite sum of positive terms can yield a finite result was viewed both as a deep philosophical challenge and an important gap in the understanding of infinity. Infinite Series were used throughout the development of the calculus and it is thus difficult to trace their exact historical path. However, there were several problems that involved infinite series that were of significant historical importance. This section contains selected problems that represent an introduction to the historical significance of the Infinite Series.

James Gregory's Infinite Series for arctan

Most of Gregory's work was expressed geometrically, and was difficult to follow. He had all the fundamental elements needed to develop calculus by the end of 1668, but lacked a rigorous formulation of his ideas. The discovery of the infinite series for arctan x is attributed to

James Gregory, though he also discovered the series for $\tan x$ and $\sec x$.

Here is how one can find the derivative of $\arctan x$

$$y = \arctan x$$
$$\tan y = \tan(\arctan x)$$
$$\tan y = x$$
$$\frac{dy}{dx} \sec^2 y = 1$$
$$\frac{dy}{dx} = \frac{1}{\sec^2 y} = \frac{1}{\tan^2 y + 1}$$
$$\frac{dy}{dx} = \frac{1}{x^2 + 1} = \frac{1}{1 + x^2}$$

The above is a modern proof, Gregory used the derivative of arctan from the work of others. The infinite series for $\frac{1}{1 + x^2}$ can be found by using long division

$$\frac{1}{1 + x^2} = 1 - x^2 + x^4 - x^6 + \cdots$$

Integrating this infinite series term-by-term produces

$$\arctan x = x - \frac{x^3}{3} + \frac{x^5}{5} - \frac{x^7}{7} + \cdots$$

which is the infinite series for arctan.

Prior to Leibniz and Newton's formulation of the formal methods of the calculus, Gregory already had a solid understanding of the differential and integral, which is shown here. Although the solution above is in modern notation, Gregory was able to solve this problem with his own methods. Gregory was one of the first to relate trigonometric functions to their infinite series using calculus, although he is primarily only remembered noted for finding the infinite series for the inverse tangent.

Leibniz's Early Infinite Series

One of Leibniz's earlier experiences with infinite series was to find the sum of the reciprocals of the triangular numbers, or $\dfrac{2}{n(n+1)}$.

By using partial fraction decomposition, the fraction can be split so that

$$\frac{2}{n(n+1)} = 2\left(\frac{1}{n} - \frac{1}{n+1}\right)$$

The first n terms of the series are

$$2\left(\frac{1}{1} - \frac{1}{1+1}\right) + 2\left(\frac{1}{2} - \frac{1}{2+1}\right) + 2\left(\frac{1}{3} - \frac{1}{3+1}\right) + \cdots + 2\left(\frac{1}{n} - \frac{1}{n+1}\right)$$

By factoring out the 2 and by rearranging the terms

$$2\left(\frac{1}{1} + \frac{1}{2} + \frac{1}{3} + \cdots + \frac{1}{n} - \frac{1}{1+1} - \frac{1}{2+1} - \cdots - \frac{1}{n+1}\right)$$

all but the first and last term cancel, and the sum reduces to $2\left(\dfrac{1}{1} - \dfrac{1}{n+1}\right)$ since

$$\lim_{n \to \infty} 2\left(\frac{1}{1} - \frac{1}{n+1}\right) = 2$$

Therefore, the sum of the reciprocals of the triangular numbers is 2.

This problem was historically significant as it served as in inspiration for Leibniz to explore many more infinite series. Since he successfully solved this problem, he concluded that a sum could be found of almost any infinite series.

Leibniz and the Infinite Series for Trigonometric Functions

After having already developed methods for differentiation and integration, Leibniz was able to find an infinite series for $\sin(z)$ and $\cos(z)$. He began the process by starting with the equation for a unit circle

$$x^2 + y^2 = 1 \quad \text{where} \quad x = \cos\theta, \, y = \sin\theta$$

and differentiating with respect to x

$$\frac{d\theta}{dx} = -\frac{1}{\sin\theta}$$

$$d\theta = -\frac{1}{\sin\theta}dx$$

By the equation of the unit circle given above $\sin\theta = \sqrt{1-\cos^2\theta}$ and $\cos\theta = x$ so

$$d\theta = -\frac{1}{\sqrt{1-x^2}}dx$$

Prior to Leibniz attempting to solve this problem, Newton had discovered the binomial theorem. Therefore, by simple application of Newton's rule, Leibniz was able to expand the equation into an infinite series

$$-\frac{1}{\sqrt{1-x^2}}dx = -\left(1 + \frac{x^2}{2} + \frac{3x^4}{8} + \frac{5x^6}{16} + \cdots\right)dx$$

Leibniz then integrated both sides. The right side of the equation can be integrated term-by-term and the left side of the equation is equal to $\arcsin(x)$. This can easily be shown

$$y = \arcsin x$$
$$\sin y = \sin(\arcsin x)$$
$$\sin y = x$$
$$\cos y \frac{dy}{dx} = 1$$
$$\frac{dy}{dx} = \frac{1}{\cos y} = \frac{1}{\sqrt{1-\sin^2 y}} = \frac{1}{\sqrt{1-x^2}}$$

Therefore, integrating both sides yields

$$\arcsin x = x + \frac{x^3}{6} + \frac{3x^5}{40} + \frac{5x^7}{112} + \cdots$$

At this point, Leibniz had found the infinite series for $\arcsin(x)$, a result which Newton had found as well. Leibniz then used a process he and Newton both discovered independently: Series Reversion. That is, given the infinite series for a function, he found a way to calculate the infinite series for the inverse function.

In this case, the process worked by first taking the sin (the inverse function for arcsin) of both sides of the equation

$$y = \arcsin x = x + \frac{x^3}{6} + \frac{3x^5}{40} + \frac{5x^7}{112} + \cdots$$

$$\sin y = \sin(\arcsin x) = \sin\left(x + \frac{x^3}{6} + \frac{3x^5}{40} + \frac{5x^7}{112} + \cdots\right)$$

$$\sin y = x = \sin\left(x + \frac{x^3}{6} + \frac{3x^5}{40} + \frac{5x^7}{112} + \cdots\right)$$

Now Leibniz assumed that an infinite series for $\sin(y)$ exists that is of the form

$$\sin y = a_1 y^1 + a_2 y^2 + a_3 y^3 + \cdots + a_n y^n + \cdots$$

Leibniz had said that $\sin y = x$, therefore for each instance of x in

$$\sin\left(x + \frac{x^3}{6} + \frac{3x^5}{40} + \frac{5x^7}{112} + \cdots\right)$$

he substituted the assumed infinite series for $\sin y$. He knew that the result of substituting in the this series for x must yield y, as it was stated

$$\sin y = \sin\left(x + \frac{x^3}{6} + \frac{3x^5}{40} + \frac{5x^7}{112} + \cdots\right)$$

Therefore, he knew the coefficient of the first term $a_1 = 1$ and all of the other coefficients must add up to 0. In order to further explain, the first 3 coefficients of the expansion will be solved for. When the series is substituted, the only possible way to have a term is when it is substituted for x. The first term in the expansion will therefore be $(a_1 1) y^1$. There will be no y^2 term, and the y^3 term will be obtained by both the 3rd power y term being plugged into x, and the 1st power y term being plugged into x^3 thus yielding $(a_1^3 \frac{1}{6} y^3 + a_3 1 y^3)$ where the sum of the coefficients must be 0 (because there is no y^3 term left over in the expansion). The same process yields the equation for the 5th order term which is $(a_5 1 y^5 + a_1^5 \frac{3}{40} y^5 + 3 a_1^2 a_3 \frac{1}{6} y^5)$. At this point the resulting expansion is

$$(a_1 1)y^1 + \left(a_1^3 \frac{1}{6} + a_3 1\right)y^3 + \left(a_5 1 + a_1^5 \frac{3}{40} + 3a_1^2 a_3 \frac{1}{6}\right)y^5 + \cdots = y$$

Now equations for each coefficient can be set up

$$a_1 = 1$$

$$a_1^3 \frac{1}{6} + a_3 1 = 0$$

$$a_5 1 + a_1^5 \frac{3}{40} + 3a_1^2 a_3 \frac{1}{6} = 0$$

Solving the equations using the previous results in each calculation yields

$$a_1 = 1 = \frac{1}{1!}$$

$$a_3 = -\frac{1}{6} = -\frac{1}{3!}$$

$$a_5 = \frac{1}{120} = \frac{1}{5!}$$

Substituting the coefficients back into the assumed infinite series for $\sin y$, he determined that

$$\sin y = y - \frac{1}{3!}y^3 + \frac{1}{5!}y^5 + \cdots$$

By simply differentiating this equation term-by-term Leibniz was also able to find the infinite series for $\cos y$.

Leibniz not only laid the groundwork for the Taylor series, but he (and simultaneously Newton) was the first to discover the series for these trigonometric functions. He invented his own method for finding the infinite series of a function's inverse.

Euler's Sum of the Reciprocals of the Squares of the Natural Numbers

Much work was done with infinite series by Euler. He was able to use infinite series to solve problems that other mathematicians were not able to solve by any methods. Neither Leibniz nor Jacques Bernoulli were able to find the sum of the inverse of the squares-they even admitted as much. The sum was unknown until Euler found it through the manipulation of an infinite series

$$\frac{1}{1^2} + \frac{1}{2^2} + \frac{1}{3^2} + \cdots$$

In order to find this sum, Euler started by examining the infinite series for sin z

$$\sin z = z - \frac{z^3}{3!} + \frac{z^5}{5!} - \frac{z^7}{7!} + \cdots$$

Equating sin z to zero gave Euler the roots of the infinite expansion

$$\sin z = 0 = z - \frac{z^3}{3!} + \frac{z^5}{5!} - \frac{z^7}{7!} + \cdots$$

That is, the roots of this equation are $z = \pi, 2\pi, 3\pi, 4\pi, \cdots$ Now, left with the equation $0 = z - \frac{z^3}{3!} + \frac{z^5}{5!} - \frac{z^7}{7!} + \cdots$ with roots $z = \pi, 2\pi, 3\pi, 4\pi, \cdots$ dividing by z results in

$$0 = 1 - \frac{z^2}{3!} + \frac{z^4}{5!} - \frac{z^6}{7!} + \cdots$$

substituting $z^2 = \omega$ yields

$$0 = 1 - \frac{\omega^1}{3!} + \frac{\omega^2}{5!} - \frac{\omega^3}{7!} + \cdots$$

and

$$\omega = (\pi)^2, (2\pi)^2, (3\pi)^2, (4\pi)^2, \cdots$$

By using properties involving polynomials, it is known that the sum of the reciprocals of the roots is the negative of the coefficient of the linear term, assuming the constant term is 1. Applying this here, we get

$$\frac{1}{6} = \frac{1}{(\pi)^2} + \frac{1}{(2\pi)^2} + \frac{1}{(3\pi)^2} + \frac{1}{(4\pi)^2} + \cdots$$

multiplying through by π^2, we get

$$\frac{\pi^2}{6} = \frac{1}{1^2} + \frac{1}{2^2} + \frac{1}{3^2} + \frac{1}{4^2} + \cdots$$

Which is the sum of the inverse of the squares. Starting with cos x instead of sin x, he obtained the sum for the sum of the squares of the odd natural numbers. He solved problems using infinite series that could not be done in any other way, and developed new ways to manipulate them.

Words and Expressions

infinite series 无穷级数
arctan = arc tangent 反正切
trigonometric *adj*. 三角的,三角学的,三角法的
trigonometric functions 三角函数,圆函数
inverse tangent 反正切
reciprocal *n*. 倒数 *adj*. 倒数的,互反的,互逆的
partial *adj*. 偏的;部分的
partial fraction 部分分式;部分分数
decomposition *n*. 分解
unit circle 单位圆
binomial *n*. 二项式 *adj*. 二项的,二项式的
the binomial theorem 二项式定理
reversion *n*. 反回
Taylor series 泰勒奇数
inverse function 反函数
manipulation *n*. 操作

5.3 DIFFERENTIAL EQUATION

A differential equation is an equation containing one or more derivatives or differentials. If there are no derivatives higher than first order in the equation, it is a first order differential equation. More precisely, a differential equation of the first order is an equation of the type

$$F(x, y, y') = 0 \tag{1}$$

Where y' is the derivative of y with respect to x.

A solution of a differential equation is a relation, free of derivatives, which satisfies it. If the variables involved are x and y, the solution may be written in the form $F(x,y) = 0$.

An equation such as (1) is really not new to the student who has studied differential calculus. Suppose for example, that we are required to find the equation of the tangent line to the circle $x^2 + y^2 = 25$ at the point $(3,4)$. We find by differentiating that the slope of the tangent at any point on the circle is

$$\frac{dy}{dx} = -x/y$$

and at the point $(3,4)$ is

$$\frac{dy}{dx} = -\frac{3}{4} \qquad (2)$$

Equation (2) is a differential equation. Moreover, the student can readily solve it to obtain

$$Y = -\frac{3}{4}x + C \qquad (3)$$

It is easy to verify that the relation (3) satisfies equation (2) for every value of C; that is, equation (2) has infinitely many solutions. There are the family of parellel lines with slope $-\frac{3}{4}$. The line we seek is

$$y = -\frac{3}{4}x + \frac{25}{4} \qquad (4)$$

which passes through the point $(3,4)$. Equation (3) identifies the general solution of the differential equation (2). Equation (4) gives us a particular solution.

If the differential equation contains a derivative of the second order but none higher, then it is called a differential equation of the second order. An example would be the differential equation

$$\frac{d^2y}{dx^2} = x \qquad (5)$$

This equation can be solved by integrating once to obtain

$$\frac{dy}{dx} = \frac{x^2}{2} + C_1$$

and integrating again to have

$$y = \frac{x^3}{6} + C_1 x + C_2 \qquad (6)$$

The solution of (5), represented by (6), contains two arbitrary constants. A solution of a second-order differential equation which contains two essential arbitrary constants is called the general solution.

A differential equation of the nth order is an equation of the type

$$F(x, y, y' \cdots y^{(n)}) = 0 \qquad (7)$$

Where $y', y'', \cdots, y^{(n)}$ represent the first, second $\cdots n$th order derivatives of y with respect to x. A solution of equation (7) which contains n essential arbitrary constants is called the general solution. Particular solutions may be obtained from the general solution by assigning values to one or more of the arbitrary contants.

The arbitrary contants are "essential" if they are not reducible in number by a mere change of notation. For instance, $y = (C_1 + C_2)x$ is not the general solution of a differential equation of second order, for, although there are apparently two arbitrary constants, C_1 and C_2, the sum $C_1 + C_2$ is also an arbitary constants and can be denoted by K. By a change of notation the solution reduces to $y = Kx$, which contains only one arbitrary constant and is the general solution of a differential equation of first order. Likewise $ax + by + c = 0$ is not the general solution of a differential equation of third order for the arbitrary constants a, b, c are not all essential. Dividing through by one of them, say c, and writing $\frac{a}{c} = A, \frac{b}{c} = B$, reduces the equation to $Ax + By + 1 = 0$, which contains two essential arbitrary constants and is the general solution of a differential equation of second order.

The existence of a differential equation does not imply the existence of any solution of the equation. The question of existence of solutions is complicated one.

All the differential equations mentioned above are ordinary differential equations. The word "ordinary" is used to distinguish them from partial differential equations, which are differential equations containing partial derivatives.

Applications of differential equations are numerous in mathematics, the natural sciences, and social studies. It is therefore useful to be able to solve them. It seems natural to solve a differential equation by integrating.

Words and Expressions

differential equation 微分方程
first order differential equation 一阶微分方程
precisely *adv*. 精确地;严格地
parallel lines 平行线
slope *n*. 斜率
arbitrary *n*. 任意的,随意的
arbitrary constant 任意常数
essential *n*. 必要的;本质的
general solution 通解
particular solution 特解
assign *vt*. 指定;分配
assigning value 指定值
reducible *n*. 可归约的,可简化的
ordinary *adj*. 常的,寻常的,正常的
ordinary differential equation 常微分方程
partial *n*. 偏的,部分的
partial differential equation 偏微分方程

5.4 FERMAT'S LAST THEOREM

A theorem first proposed by Fermat in the form of a note scribbled in the margin of his copy of the ancient Greek text *Arithmetica* by Diophantus. The scribbled note was discovered posthumously, and the original is now lost. However, a copy was preserved in a book published by Fermat's son. In the note, Fermat claimed to have discovered a proof that the Diophantine equation $x^n + y^n = z^n$ has no integer solutions for $n > 2$ and $x, y, z \neq 0$.

The full text of Fermat's statement, written in Latin, reads "Cubum autem in duos cubos, aut quadrato-quadratum in duos quadrato-quadratos, et generaliter nullam in infinitum ultra quadratum potestatem in duos eiusdem nominis fas est dividere cuius rei demonstrationem mirabilem sane detexi. Hanc marginis exiguitas non caperet". In translation, "It is impossible for a cube to be the sum of two cubes, a fourth power to be the sum of two fourth powers, or in general for any number that is a power greater than the second to be the sum of two like powers. I have discovered a truly marvelous demonstration of this proposition that this margin is too narrow to contain."

As a result of Fermat's marginal note, the proposition that the Diophantine equation

$$x^n + y^n = z^n \qquad (1)$$

where x, y, z, and n are integers, has no nonzero solutions for $n > 2$ has come to be known as Fermat's Last Theorem. It was called a "theorem" on the strength of Fermat's statement, despite the fact that no other mathematician was able to prove it for hundreds of years.

Note that the restriction $n > 2$ is obviously necessary since there are a number of elementary formulas for generating an infinite number of Pythagorean triples (x, y, z) satisfying the equation for $n = 2$

$$x^2 + y^2 = z^2 \qquad (2)$$

A first attempt to solve the equation can be made by attempting to factor the equation, giving

$$(z^{\frac{n}{2}} + y^{\frac{n}{2}})(z^{\frac{n}{2}} - y^{\frac{n}{2}}) = x^n \tag{3}$$

Since the product is an exact power

$$\begin{cases} z^{\frac{n}{2}} + y^{\frac{n}{2}} = 2^{n-1}p^n \\ z^{\frac{n}{2}} - y^{\frac{n}{2}} = 2q^n \end{cases} \text{ or } \begin{cases} z^{\frac{n}{2}} + y^{\frac{n}{2}} = 2p^n \\ z^{\frac{n}{2}} - y^{\frac{n}{2}} = 2^{n-1}q^n \end{cases} \tag{4}$$

Solving for y and z gives

$$\begin{cases} z^{\frac{n}{2}} = 2^{n-2}p^n + q^n \\ y^{\frac{n}{2}} = 2^{n-2}p^n - q^n \end{cases} \text{ or } \begin{cases} z^{\frac{n}{2}} = p^n + 2^{n-2}q^n \\ y^{\frac{n}{2}} = p^n - 2^{n-2}q^n \end{cases} \tag{5}$$

which give

$$\begin{cases} z = (2^{n-2}p^n + q^n)^{\frac{2}{n}} \\ y = (2^{n-2}p^n - q^n)^{\frac{2}{n}} \end{cases} \text{ or } \begin{cases} z = (p^n + 2^{n-2}q^n)^{\frac{2}{n}} \\ y = (p^n - 2^{n-2}q^n)^{\frac{2}{n}} \end{cases} \tag{6}$$

However, since solutions to these equations in rational numbers are no easier to find than solutions to the original equation, this approach unfortunately does not provide any additional insight.

It is sufficient to prove Fermat's Last Theorem by considering prime powers only, since the arguments can otherwise be written

$$(x^m)^p + (y^m)^p = (z^m)^p \tag{7}$$

so redefining the arguments gives

$$x^p + y^p = z^p \tag{8}$$

The so-called "first case" of the theorem is for exponents which are relatively prime to x, y, and z ($p + x, y, z$) and was considered by Wieferich. Sophie Germain proved the first case of Fermat's Last Theorem for any odd prime p when $2p + 1$ is also a prime. Legendre subsequently proved that if p is a prime such that $4p + 1, 8p + 1, 10p + 1, 14p + 1$, or $16p + 1$ is also a prime, then the first case of Fermat's Last Theorem holds for p. This established Fermat's Last Theorem for $p < 100$. In 1849, Kummer proved it for all regular primes and composite numbers of which they are factors.

Kummer's attack led to the theory of ideals, and Vandiver developed Vandiver's criteria for deciding if a given irregular prime satisfies the theorem. Genocchi (1852) proved that the first case is true for p if $(p, p-3)$ is not an irregular pair. In 1858, Kummer showed that the first case is true if either $(p, p-3)$ or $(p, p-5)$ is an irregular pair, which was subsequently extended to include $(p, p-7)$ and $(p, p-9)$ by Mirimanoff (1905). Vandiver pointed out gaps and errors in Kummer's memoir which, in his view, invalidate Kummer's proof of Fermat's Last Theorem for the irregular primes 37, 59, and 67, although he claims Mirimanoff's proof of FLT for exponent 37 is still valid.

Wieferich (1909) proved that if the equation is solved in integers relatively prime to an odd prime p, then
$$2^{p-1} \equiv 1 \pmod{p^2} \qquad (9)$$
(Ball and Coxeter 1987). Such numbers are called Wieferich primes. Mirimanoff (1909) subsequently showed that
$$3^{p-1} \equiv 1 \pmod{p^2} \qquad (10)$$
must also hold for solutions relatively prime to an odd prime p, which excludes the first two Wieferich primes 1093 and 3511. Vandiver (1914) showed
$$5^{p-1} \equiv 1 \pmod{p^2} \qquad (11)$$
and Frobenius extended this to
$$11^{p-1}, 17^{p-1} \equiv 1 \pmod{p^2} \qquad (12)$$
It has also been shown that if p were a prime of the form , then
$$7^{p-1}, 13^{p-1}, 19^{p-1} \equiv 1 \pmod{p^2} \qquad (13)$$
which raised the smallest possible p in the "first case" to 253, 747, 889 by 1941 (Rosser 1941). Granville and Monagan (1988) showed if there exists a prime p satisfying Fermat's Last Theorem, then
$$q^{p-1} \equiv 1 \pmod{p^2} \qquad (14)$$
for $q = 5, 7, 11, \cdots, 71$. This establishes that the first case is true for all

prime exponents up to 714, 591, 416, 091, 398 (Vardi 1991).

The "second case" of Fermat's Last Theorem (for $p \mid x, y, z$) proved harder than the first case.

Euler proved the general case of the theorem for $n = 3$, Fermat $n = 4$, Dirichlet and Lagrange $n = 5$. In 1832, Dirichlet established the case $n = 14$. The $n = 7$ case was proved by Lamé, using the identity

$$(X + Y + Z)^7 - (X^7 + Y^7 + Z^7) = 7(X + Y)(X + Z)(Y + Z) \times$$
$$[(X^2 + Y^2 + Z^2 + XY + XZ + YZ)^2 + XYZ(X + Y + Z)] \quad (15)$$

Although some errors were present in this proof, these were subsequently fixed by Lebesgue (1840). Much additional progress was made over the next 150 years, but no completely general result had been obtained. Buoyed by false confidence after his proof that π is transcendental, the mathematician Lindemann proceeded to publish several proofs of Fermat's Last Theorem, all of them invalid. A prize of 100 000 German marks, known as the Wolfskehl Prize, was also offered for the first valid proof.

A recent false alarm for a general proof was raised by Y. Miyaoka (Cipra 1988) whose proof, however, turned out to be flawed. Other attempted proofs among both professional and amateur mathematicians are discussed by vos Savant (1993), although vos Savant erroneously claims that work on the problem by Wiles (discussed below) is invalid. By the time 1993 rolled around, the general case of Fermat's Last Theorem had been shown to be true for all exponents up to 4×10^6 (Cipra 1993). However, given that a proof of Fermat's Last Theorem requires truth for *all* exponents, proof for any finite number of exponents does not constitute any significant progress towards a proof of the general theorem.

In 1993, a bombshell was dropped. In that year, the general theorem was partially proven by Andrew Wiles by proving the semistable case of the Taniyama-Shimura conjecture. Unfortunately, several holes were discovered in the proof shortly thereafter when Wiles' approach via the Taniyama-Shimura conjecture became hung up on properties of the

Selmer group using a tool called an Euler system. However, the difficulty was circumvented by Wiles and R. Taylor in late 1994 and published in Taylor and Wiles (1995) and Wiles (1995). Wiles' proof succeeds by (1) replacing elliptic curves with Galois representations, (2) reducing the problem to a class number formula, (3) proving that formula, and (4) tying up loose ends that arise because the formalisms fail in the simplest degenerate cases.

The proof of Fermat's Last Theorem marks the end of a mathematical era. Since virtually all of the tools which were eventually brought to bear on the problem had yet to be invented in the time of Fermat, it is interesting to speculate about whether he actually was in possession of an elementary proof of the theorem. Judging by the temerity with which the problem resisted attack for so long, Fermat's alleged proof seems likely to have been illusionary. This conclusion is further supported by the fact that Fermat searched for proofs for the cases $n = 4$ and $n = 5$, which would have been superfluous had he actually been in possession of a general proof.

Words and Expressions

the Diophantine equation 丢番图方程
integer solution 积分解
cube n. 立方,立方体
marvelous adj. 奇异的,不平常的,了不起的
demonstration n. 证明,示教
nonzero adj. 非零的
nonzero solution 非零解
Pythagorean n. 毕达哥拉斯学派 adj. 毕达哥拉斯的
triply adv. 三重地,三倍地
factor adj. 因式,因子,因数

5.5 REVEALING GOLDBACH CONJECTURE

5.5.1 Put Forward a Question

It is a simple problem that an even number can decompose the sum of two prime numbers. But whether it is true for all the even numbers is a complicated problem. In recent two hundred years, this problem is always disturbing the mathematics field and we have not a satisfying answer. In our spare time, we studied this problem and we thought we have got the answer. Now we bring it forward and we hope to get a pertinent comment.

To this conjecture, we think it is true. Why? Let us look through these theorems, formulas, lists and some proofs.

5.5.2 The Infinity Theorem of Prime Number

Theorem: Prime numbers are infinite in natural numbers .

For this theorem has been a law, we need not to prove it. But it is a key position in the proof of the conjecture. Because of this theorem, we can know prime number's distribution in natural numbers are prolonging with the natural numbers table's extension. Prime number is endless like natural. In fact, prime numbers are natural numbers. We take this theorem out in order to resolve this problem conveniently.

If a prime number is limited, it must have a maximum element N. So to an even number which is equal to and more than 2N. It is impossible to find one pair or some pairs of prime numbers to decompose this even number. They exist in natural numbers list endlessly. Only by this, the conjecture's truth can be ensured.

5.5.3 List of Prime Numbers Within 10 000

In order to prove the conjecture, first we should make a prime number list in 10 000. This list is different from general list. We put every number in its position respectively, so it is very distinct and easy to use (referring to another graph the prime numbers list in 10 000).

The making way of the prime numbers list of four column is prime the method of the prime number's radiation, which means: from the least prime number 2, if we radiate the prime number 2,3,5, 7,11,13,⋯ in order, after subtracting each radiation numbers from natural list. The remained numbers in the list are prime numbers. These prime numbers aspect 2 and 5, all aggregate in column B1, B3, B7, B9 in natural numbers list. For in the form of "four columns" we call this list "prime numbers list in 10 000 in four columns". Here is the making process.

a. First we set up a natural numbers list in 10 000

We use "A" to make the rows of this list. The numerals are determined by the natural numbers that are in the places more than ten and we also refer to unit's place. When the data ten's place is "0", the numerals of "A" need not to be marked. When it is not "0", (i.e. 1,2, 3,4,5,6,7,8,9), the numerals of "A" adds one. For example, 320 is located in the row of A_{32}, and 321, 322, ⋯, 329 are located in the row of A_{33}.

We use "B" to make the columns of the list. The numerals are determined by the datum in unit's place, For example, 1,11,21,⋯ are located in B_1; 2,12,22,⋯ are located in B_2; 3,13,23,⋯ are located in B_3. It's the same to others.

Graph 1   List of prime number within 10 000

	B_1	B_2	B_3	B_4	B_5	B_6	B_7	B_8	B_9	B_0
A_1	1	2	3	4	5	6	7	8	9	10
A_2	11	12	13	14	15	16	17	18	19	20
A_3	21	22	23	24	25	26	27	28	29	30
A_4	31	32	33	34	35	36	37	38	39	40
A_5	41	42	43	44	45	46	47	48	49	50

	B_1	B_2	B_3	B_4	B_5	B_6	B_7	B_8	B_9	B_0
A_6	51	52	53	54	55	56	57	58	59	60
A_7	61	62	63	64	65	66	67	68	69	70
A_8	71	72	73	74	75	76	77	78	79	80
A_9	81	82	83	84	85	86	87	88	89	90
A_{10}	91	92	93	94	95	96	97	98	99	100
A_{11}	101	102	103	104	105	106	107	108	109	110
A_{12}	111	112	113	114	115	116	117	118	119	120
A_n		...	...							

b. Prime number's radiation and radiation number

"1" in natural number is unit number, so we remove it from the list. The following number is "2" which is a prime number. Multiply 2 by the number which is equal to 2 and more than 2. i.e. $2 \times 2 = 4$, $2 \times 3 = 6$, $2 \times 4 = 8$, $2 \times 5 = 10$, $2 \times 6 = 12$, $\cdots$, these products are radiation numbers of 2. If we remove these products from our list, we still get the radiation numbers list of prime number "2". When we remove these radiation numbers of 2 from the natural number list, the data in column B_2, B_4, B_6, B_8, B_0 are removed except 2 (referring to graph 2).

Graph 2　The radiation numbers list of prime number 2

	B_1	B_3	B_5	B_7	B_9
A_1	2	3	5	7	9
A_2	11	13	15	17	19
A_3	21	23	25	27	29
A_4	31	33	35	37	39
A_5	41	43	45	47	49
A_6	51	53	55	57	59
A_7	61	63	65	67	69
A_8	71	73	75	77	79
A_9	81	83	85	87	89

	B_1	B_3	B_5	B_7	B_9
A_{10}	91	93	95	97	99
A_{11}	101	103	105	107	109
A_{12}	111	113	115	117	119
A_n	...	...			

The next prime number is 3. Its radiation number is remained numbers. When we remove the product of multiply 3 by the number which is equal to 3 and more than 3 from Graph 2, i.e. $3 \times 3 = 9$, $3 \times 5 = 15$, $3 \times 7 = 21$, $3 \times 9 = 27$, $3 \times 11 = 33$, $3 \times 13 = 39$, $3 \times 15 = 45$, $3 \times 17 = 51$, $3 \times 19 = 57$, ..., $3 \times 3\ 333 = 9\ 999$. If we remove these products from Graph 2, we'll get Graph 3.

Graph 3

	B_1	B_3	B_5	B_7	B_9
A_1	2	3	5	7	
A_2	11	13		17	19
A_3		23	25		29
A_4	31		35	37	
A_5	41	43		47	49
A_6		53	55		59
A_7	61		65	67	
A_8	71	73		77	79
A_9		83	85		89
A_{10}	91		95	97	
A_n	...	...			

The radiation numbers of 5 are: $5 \times 5 = 25$, $5 \times 7 = 35$, $5 \times 11 = 55$, $5 \times 13 = 65$, $5 \times 17 = 85$, $5 \times 19 = 95$, ..., $5 \times 1\ 999 = 9\ 995$. Remove these products from Graph 3, we'll get Graph 4.

Graph 4

	B_1	B_3	B_7	B_9
A_1	2	3	7	
A_2	11	13	17	19
A_3		23		29
A_4	31		37	
A_5	41	43	47	49
A_6		53		59
A_7	61		67	
A_8	71	73	77	79
A_n	...	...		

The radiation numbers of 7 are $7 \times 7 = 49$, $7 \times 11 = 77$, $7 \times 13 = 91$, $7 \times 17 = 119$, $7 \times 19 = 133$, $7 \times 23 = 161$, $7 \times 29 = 203$, $\cdots$, $7 \times 1\,427 = 9\,989$. Remove these products from Graph 4, we'll get Graph 5.

Graph 5

	B_1	B_3	B_7	B_9
A_1	2	3	7	
A_2	11	13	17	19
A_3		23		29
A_4	31		37	
A_5	41	43	47	
A_6		53		59
A_7	61		67	
A_8	71	73		79
A_9		83		89
A_{10}			97	
A_{11}	101	103	107	109
A_{12}		113		
A_n	...	...		

The following prime numbers are 11, 13, 17, 19, 23, 29, 31, 37, 41, 43, 47, 53, 59, 61, 67, 71, 73, 79, 83, 89, 97. If we radiate these 25 prime numbers in order and remove the radiation numbers from the radiation number list, we'll get a number list on which are all prime numbers. Now we set up a prime number list in 10 000.

c. Differentiation between prime number district and nonprime number district

Prime number district is the following: when we don't radiate a prime number, the numbers which are less than the square number of this prime number are all prime numbers, which is the district of the prime number.

For example, to prime number "2", $2 \times 2 = 4$, the prime numbers which are less than 4 and are in the radiation numbers list are 2, 3, then this district is prime number district of "3".

To prime number "3", $3 \times 3 = 9$, the prime numbers which are less than 9 and are in the radiation numbers list are 5, 7, then this district is prime number district of "3".

To prime number "5", $5 \times 5 = 25$, the numbers which are less than 25 and are in the radiation number list are 11, 13, 17, 19, 23, then this district are prime numbers district of "5".

To prime number "7", $7 \times 7 = 49$, the numbers which are less than 49 and are in the radiation number list are 29, 31, 37, 41, 43, 47, then this district are prime numbers district of "7".

To the other prime numbers such as 11, 13, 17, 19, 23, 29, 31, 37, 41, 43, 47, 53, 59, 61, 67, 71, 73, 79, 83, 89, 97, the prime numbers district is amplifying these prime numbers until we radiate all of these prime numbers. If we remove radiation numbers from radiation number list, we will get a prime number list in 10 000.

Words and Expressions

conjecture *n.* 猜想,推测
even number 偶数
decompose *n.* 分解,腐烂 *vt.* 分解
pertinent *adj.* 适当的,恰当的,中肯的
comment *n.* 注解,备注,注释,注解,注解 *vi.* 注释 *vt.* 注解,评论
prolong *n.* 持续
radiation *n.* 放射,辐射能,辐射,射线 *vi.* 辐射
aggregate *vi.* 集合,总计 *adj.* 总计的,聚生的,聚集的 *n.* 综合指标,集合物,复合机组,骨料,聚集 *vt.* 总计,使聚集,法人团体
numerals *n.* 数字,数码
datum *n.* 资料,数据,已知数,材料,已知件
differentiation *n.* 分化,微分法,区别,区别,分化作用
district *n.* 管区,地域,区,地区,区域
amplifying *vt.* 扩大,放大,增大
theorem *n.* 定理,法则,一般原则

6

Others

6.1 MATHEMATICS IN TODAY'S FINANCIAL MARKETS

Market—this idea is usually associated with institutions, people, and actions involved in trading valuables. The valuables, or assets, are called securities (soon we shall talk about these in more detail). The place where we trade them is called a financial market. Not only people, but banks, firms, investment and insurance companies, pension funds, and other structures participate in financial markets.

Buying and selling, owning and loaning assets, receiving dividends, and consuming capital are some of the activities that take place at financial markets. In modern days, these activities require serious quantitative calculations, which we cannot conduct unless we "idealize" the market. For instance, we must assume that all operations and transactions take place immediately (this is called a liquid market) and that they are free (the notion of a frictionless market). Securities are the basis for a financial market. Securities come in many shapes and kinds, but the main ones are stocks and bonds.

Stocks are securities that hold a share of the value of the company (the words stocks and shares are used interchangeably). A company issues stock when it needs to raise capital (money). People buy stock and thus own a "piece" of a company. This ownership gives stockholders the right to make decisions in the way the company is governed and to receive dividends based on the amount of stock a person has.

Bonds are other instruments that a government or a company issues when it needs to raise money. In effect, the buyers of bonds lend their money to the institutions that issue bonds. But such debt must be paid off, and this is done in two ways. Unlike stocks, bonds have an expiry date, indicating when the original borrowed amount (nominal value or principal value) must be paid to the lender. In addition, throughout the term of the bond, the lender receives coupon payments according to the "yield" indicated on the bond. The bond yield is a very significant quantitative indicator for financial calculations; it is similar to a bank's rate of interest—the "reward" for investing money in that bank. The bond without coupon payments can viewed as the "money" in the bank account.

Let's say that a bank persuades you to invest your funds (and now, at time 0, you have the amount B_0) in one of its accounts for a certain period of time (one month, three months, one year, etc.) by promising that at the end of this period (time 1), you will receive a risk-free yield, that is, your initial investment will increase by an amount denoted ΔB_1. Note that $\Delta B_1 = B_1 - B_0$, and also $\Delta B_1 = rB_0$, where r is the interest coeffcient, or the bank's interest rate.

Depending on whether you decide to reinvest (monthly, quarterly, yearly), you will receive only the initial investment, or that plus the interest you have earned after $n = 1, 2, 3, \cdots$ periods of time (see Figure 18)

$$\begin{cases} B_n = B_{n-1} + rB_0 = B_0(1 + rn) \\ \quad \text{or} \\ B_n = B_{n-1}(1 + r) = B_0(1 + r)^n \end{cases} \quad (1)$$

The relationship $\dfrac{\Delta B_n}{B_{n-1}} = \dfrac{(B_n - B_{n-1})}{B_{n-1}} = r$ characterizes the yield of your investment.

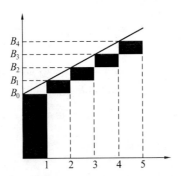

Figure 18　Simple interest-linear growth.

Usually, the rate of interest, or the "yield" on the investment, r 100%, is stated for a year. We can divide this time period into m smaller periods and calculate the yield (monthly, quarterly, semi-annually, etc.) at the end of each period, according to the stated annual rate. More frequent compounding leads to an increase in the investor's capital; the amount $B_n^{(m)}$ is given by

$$B_n^{(m)} = B_0 \left(1 + \frac{r}{m}\right)^{mn} \quad (2)$$

If we subdivide the year into more and more periods, so that m approaches infinity, then $B_n^{(m)}$ approaches $B_0 e^n$. In other words, the "limiting" amount of money in the bank account is

$$\lim_{m \to \infty} B_n^{(m)} = B_0 e^n \quad (3)$$

This implies that the relative yield of such an investment is constant and equals the interest rate r.

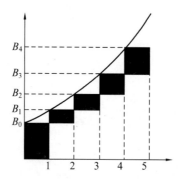

Figure 1 Compound interest-non-linear (exponential) growth.

The three methods of calculating interest discussed above are called simple, compound, and continuous. Formulas (1),(2) and (3) provide ways to calculate the amount in the investor's bank account and clearly show the dependence of the value of money on time.

On the other hand, we have the bank's interest rate R, so that if we invest the amount $(1 - R)B_1$, at time 1, say, after one year, we will receive the amount B_1. This is equivalent to the issuance of a bond with a nominal value B_1 (to be paid to the bond holder at the end of this year), but now the bond sells for a lower price (decreased by the amount of the lending rate **R** over one period, that is, $m = 1$). So today's price is determined by the formula $(1 - R)B_1$, which is equal to the discounted price $\dfrac{B_1}{(1 + r)}$. Therefore, we can view the bank account as a coupon-free bond in the sense of a risk-free asset of the financial market. The lack of, or very small, changes in interest rates characterize the stability of financial and economic systems, for which the corresponding bank account serves as the basic non-risky asset. Reality shows that such suggestions present limits in the idealization of mathematical models for financial markets.

Formulas (1), (2) and (3) show time evolution of the value of money, presenting difficulties in the calculations of annuities. These are periodic payments to be made in the future (such as rent), denoted by $f_0, f_1, f_2, \cdots, f_n$, whose values we need to know today. According to the compound interest formula, we calculate the value of the kth payment as $\dfrac{f_k}{(1+r)^k}$. Thus, the cost of all future payments today is given by the sum

$$f_0 + \frac{f_1}{(1+r)^1} + \frac{f_2}{(1+r)^2} + \cdots + \frac{f_n}{(1+r)^n}$$

These and similar arithmetic calculations determining rent payments were the only functions of mathematics in finance until the middle of the 20th century.

After the risk-free bank account, the second basic element of a financial market is a stock, which is much more volatile and thus is called a risky asset. Let S_n denote the price of the stock at time n. We determine the yield of a stock during any time period by $\rho_n = \dfrac{(S_n - S_{n-1})}{S_{n-1}}$, where $n = 1, 2, \cdots$. Then stock prices satisfy this equation

$$S_n = S_{n-1}(1 + \rho_n) \tag{4}$$

Bank account balance (1), interest rate r and stock price (4), for changing yield ρ_n form the mathematical model of a financial market.

Many factors, often very difficult to determine, cause changes in stock prices S_n. We refer to these factors as randomness and call S_n (and thus ρ_n) random variables. Just like the yield of a bank account, $r = \dfrac{\Delta B_n}{B_{n-1}}$, ρ_n is the changing yield of a risky asset (stock in our example). Note that since ρ_n changes every time period (at each $n = 1, 2, \cdots$.), we can take all these different values of ρ and calculate their mean μ, and individual yield values will lie below and above the mean. As we shorten our time periods (for instance, instead of observing changes in stock prices and thus monthly or weekly yields, we record the changes hourly or even every minute), we see that the up-and-down movements of the stock's yield become more and more chaotic. The picture below shows a possibility of such limiting behaviour of yield-per-time, where discrete time periods, divided again and again, become a continuous time line.

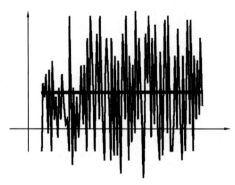

Figure 20 Varying yield values and their time mean.

Formally, in the model with continuous time, at any moment in time t, the limiting yield equals the sum

$$\mu + \sigma \widetilde{W}_t \qquad (5)$$

where μ is the mean yield, σ volatility, and $\widetilde{W}_t$ represents Gaussian "white noise," a notion used in math and physics to describe chaotic, irregular movements.

The pairs of formulas (1) and (4), and (3) and (5), respectively constitute the binomial and diffusion models of the financial market and frequently are called the Cox-Ross-Rubinstein model and the Black-Scholes model.

Further, a participant in the securities market has to invest his or her resources into assets available in this market, choosing certain quantities of different assets. We refer to this process as forming one's investment portfolio. Deciding how much of which assets to include in the investment portfolio is the essence of managing capital. Any changes to the con tents of the portfolio should limit or minimize the risk from financial operations; this is called hedging the portfolio.

Among all investment strategies, we separate those that bring profit without any initial expense. The possibility of such strategies reveals the presence of arbitrage in a financial market, which means that the market is unstable. The models we discussed above are idealized, in that they do not allow any arbitrage opportunities.

Developments in the financial market now give its participants access to instruments evolved from basic stocks and bonds. Forwards, futures, and options, called derivative securities, attract investors with lower prices. Derivative securities increase the liquidity of the market and function as insurance against losses from unsuccessful investments.

For example, consider company A, which wishes to buy stock of company B at the end of this year. The price of B's stock can either increase or decrease. So, to insure itself against higher prices, company A signs a forward contract with company B. According to the contract, A will buy B's stock at a predetermined and fixed price F at the end of the year.

Now consider another case. Say company A already has B's stock, so A wisely wants to insure itself against the losses it would incur if the price of B's stock falls. Therefore, A purchases a seller's option from B. This agreement grants A the right to sell B's stock at a predetermined and fixed price K at the end of the year. For the opportunity to do so, A pays B a price for the contract—a premium.

A future contract is similar to a forward contract, but rather than being written by the two participating sides directly, it is made through an exchange—a special organization for managing the trade of various goods, financial instruments, services, etc. At an exchange, all commercial operations are done by brokers, or intermediaries, who bring together individuals and firms to make contracts.

The first exchange specializing in the trade of options, CBOE (Chicago Board Option Exchange), opened on April 26, 1973, and by the end of the first day of work as many as 911 contracts were signed (one contract equals 100 shares). Since then, the derivative securities markets have grown fast. The huge capital of more and more participating firms and the astonishing volume of contracts being signed increases the volatility of derivative securities markets, thus increasing the randomness factor in the determination of prices of traded assets. Therefore, appropriate stochastic models have become necessary for the valuation of assets. Today, probability theory and mathematical statistics are used to develop such models for financial markets.

In the entire spectrum of securities, the most significant one, mathematically, is an option—a derivative security that gives the right to buy stock (this is a "call" option) at a predetermined price K at the termination time T. (Note that the right to sell stock is a "put" option). The exercise of a call option demands payment of $(S_T - K)^+$, which is the greater of $(S_T - K)$ and zero. Likewise, we have $(K - S_T)^+$ with a put option. The main problem, both practically and theoretically, is this: what should be the current price P_T of the contract C_T? We only need to find C_T or P_T since

$$P_T = C_T - S_0 + \frac{K}{(1+r)^T} \text{ or } P_T = C_T - S_0 + \frac{K}{e^{rT}}$$

In the case of the binomial model, we have two possibilities for the stock price at the end of a period: either it will go up with the probability p, or it will go down with the probability $1 - p$. So the stock yield ρ will take on values b in $p \cdot 100\%$ of cases and a in $(1-p) \cdot 100\%$ of cases, with $b > r > a > -1$. The exact answer for the price of the call option in the binomial model is given by the famous formula of Cox-Ross-Rubinstein (1976):

$$C_T = S_0 \sum_{k=k_0}^{T} \frac{k!}{T!(T-k)!} \tilde{p}^k (1-\tilde{p})^{T-k} - K(1+r)^{-T}$$

$$\sum_{k=k_0}^{T} \frac{k!}{T!(T-k)!} (p^*)^k (1-p^*)^{T-k}$$

where $k!$ is the product $1 \cdot 2 \cdot \cdots \cdot k$, $p^* = \frac{(r-a)}{(b-a)}$, $\tilde{p} = p^* \frac{(1+b)}{(1+r)}$, and k_0 is the smallest integer j that makes the quantity $S_0(1+a)^T \left(\frac{1+b}{1+a}\right)^j$ greater than K.

But initially, the answer was found for the diffusion model by Black, Scholes, and Merton in 1973:

$$C_T = S_0 \Phi\left(\frac{\ln\frac{S_0}{K} + T(r+\frac{\sigma^2}{2})}{\sigma\sqrt{T}}\right) - Ke^{rT}\Phi\left(\frac{\ln\frac{S_0}{K} + T(r-\frac{\sigma^2}{2})}{\sigma\sqrt{T}}\right)$$

where

$$\Phi(x) = \frac{1}{\sqrt{2\pi}} \int_{-\infty}^{x} e^{-\frac{y^2}{2}} dy$$

is the error function corresponding to the standard normal distribution. The significance of this discovery was acknowledged with the Nobel Prize in economics in 1997.

Let us remark that, historically, the first strictly mathematical work in calculating options, the Theory of Speculations, was written by L. Bachelier in 1900. However, no one saw the significance of these calculations at the time, and it was only in the middle of the 1960s famous economist Samuelson "rediscovered" Bachelier's paper and introduced the more natural market model (5), known today as the formula of Black, Scholes, and Merton.

As we have mentioned already, options and other derivative securities can function as insurance. Unlike traditional insurance, where a client "sells" his risk to some insurance company, insurance through options (hedging) allows putting this risk in the financial market with the opportunity to watch stock prices and adequately react to changes in the market situation. In this way, finance and insurance are merged. Therefore, in a financial market, the risk inherent in any investment portfolio can be managed with the insurance method described above. Insurance derivative securities (insurance forwards, futures, options), have become some of the most popular assets to be traded in the past decade. And the quantitative calculations of premiums (contract prices) and risk are done using a mix of methods in financial and actuarial mathematics.

Words and Expressions

asset *n*. 有价值的特性或技能,(常用复数)财产,资产
security *n*. 安全,保护,保障;安全措施;抵押品
participate *vi*. 参加,参与
dividend *n*. 股息,红利
quantitative *adj*. 数量的,有关数量的
idealize *vt*. 理想化
friction *n*. 摩擦,人或党派之间不同观点的矛盾或冲突
stock *n*. 股份(常用复数);公债,存货
bond *n*. 有息债券;合同;票据
expiry *n*. 终止,届满,期满,满期
coupon *n*. 证明持有人有做某事或获得某物之权利的票据
annuity *n*. 年金,养老金;年金保险
periodic *adj*. 定期的,周期的
volatile *adj*. 不稳定的,无常性的
volatility *n*. 易变,轻快;挥发,挥发性
diffusion *n*. 散步,传播,弥漫,扩散
portfolio *n*. 投资组合;公文包,文件夹

essence *n*. 本质,精髓,要素
content *n*. 知足,满足,满意,愉快
hedge *n*. 防止造成损失的手段;树篱
strategy *n*. 战略;策略,谋略;对策,政策
arbitrage *n*. 裁决,仲裁;(股票等)套利,套汇
participant *n*. 参加者
derivative *adj*. 非独创的,由他事物演变的
incur *vt*. 遭受,蒙受,招致
premium *n*. 保险费,额外费用,津贴,花红;(习语)(指公债或股票)超过正常或市面的价值,溢价
intermediary *n*. 中间人,调节人
option *n*. 可供选择的事物,选择
appropriate *adj*. 适当的,合适的,正当的
stochastic *adj*. 猜测的,好猜测的
spectrum *n*. 范围,系列;光谱
speculation *n*. 思考,思索,推断,推测
inherent *adj*. 内在的,固有的,本质的

6.2 MATHS AND MOTHS

I don't reveal to many what some might regard as my somewhat eccentric hobby of rearing caterpillars and photographing the moths that ultimately emerge. This is my form of relaxation after the day is done, and my mind by then is usually far from mathematics.

Yet there is a moth, the Peppered Moth (Biston betularia), that lends itself well to mathematical analysis. It is common in Europe and in North America, including the west coast of Canada and the United States. It is often held to represent one of the fastest known examples of Darwinian evolution by variation and natural selection. A vast literature has accumulated on this moth, both by scientists and, I recently discovered, by creationists. The latter seek to disprove the hypothesis that it is an example of evolution, and their arguments do, I suppose, at least keep scientists on their toes to ensure that their evidence is compelling.

The normal form of the moth has a "peppered" appearance. When this normal form rests on a liche-ncovered tree trunk it is very difficult to see; it is well protected by its cryptic coloration. There is another form that is almost completely black—the melanic form. It is quite conspicuous when resting on a lichen-covered tree trunk, and it is at a grave selective disadvantage. The melanic form are readily snapped up by hungry birds.

In industrial areas of nineteenth century England, long before modern atmospheric pollution controls, factory chimneys belched out huge quantities of black smoke, which killed the lichens and coated the tree trunks with dirty black grime. Suddenly the "normal" form became conspicuous, and the melanic form cryptic. Within a few generations the populations of these moths changed from almost entirely "normal" to almost entirely "melanic". This is a situation that cries out for some sort of population growth analysis. We first have to understand a little about genetics and I hope that professional geneticists will forgive me if I simplify this just a little for the purpose of this article.

The colour of the moths' wings is determined by two genes, which I denote by M for melanic and n for normal. Each moth inherits one gene from each of its parents. Consequently the "genotype" of an individual moth can be one of three types: MM, Mn or nn. MM and nn are described as "homozygous", and Mn is "heterozygous". An MM moth is melanic in appearance, and an nn moth is normal. What does an Mn moth look like? Well, surprisingly, the heterozygous moth isn't intermediate in appearance; it is melanic. Because of this, we say that the M gene is dominant over the n gene; the n gene is recessive. That is why I have written M as a capital letter and n as a small letter.

(Actually the situation is rather more complicated than this, and there are indeed intermediate forms, but the purpose of this article is to illustrate some principles of mathematical analysis of natural selection, not to bog ourselves down in detail. So I'll keep the model simple and, to begin with, I'll suppose that just the two genes are involved and that one is completely dominant over the other.)

One can see now how vulnerable the M gene is in an unpolluted environment. Not only MM moths but also Mn moths are conspicuous and are easily snapped up by birds; the M gene doesn't stand a chance. But now blacken the tree trunks. MM and Mn are black and protected; the homozygous nn form is conspicuous. The n gene, however, does not disappear, because it is protected in the Mn individuals, which are black. The populations become predominantly composed of black individuals, some of which are MM and some are Mn.

Now for our first mathematics question. Suppose we have a large population, N, of moths. Each moth will have two genes that control wing colour, so there will be $2N$ such genes distributed among the N moths. Let us suppose that a fraction x of these genes are M and a fraction $1 - x$ of them are n. What fraction of the population of moths will be genotypically MM, what fraction will be Mn, and what fraction will be nn? The answers are the successive terms of the expansion of $[x + (1 - x)]^2$. That is, the fractions of MM, Mn, and nn moths in the population will be $x^2, 2x(1 - x)$, and $(1 - x)^2$. (Verify that the sum of these is 1.) Since both MM and Mn are phenotypically melanic (i.e. melanic in external appearance), the fraction of melanic moths in the population will be $x(2 - x)$ and the fraction of normal moths will be $(1 - x)^2$.

Now, according to our theory, melanic moths in a polluted environment have a selective advantage over normal moths. Can we define "selective advantage" quantitatively? Let us suppose that a generation of moths emerges from their pupae such that the gene ratio $M:n$ is $x:1 - x$, and hence that the genotype ratio $MM:Mn:nn$ is $x^2:2x(1 - x):$

$(1 - x)^2$. Let us suppose that, by the time these moths are ready to lay their eggs to produce the next generation, the number of phenotypically melanic moths has been reduced by a factor α ($0 \leqslant \alpha \leqslant 1$) and the number of phenotypically normal moths has been reduced by a factor γ ($0 \leqslant \gamma \leqslant 1$). I define the selective advantage s of the melanic moths as

$$s = \frac{\alpha - \gamma}{\alpha + \gamma} \tag{1}$$

This number lies between -1 to $+1$. If $s = -1$, the melanic form is at a severe disadvantage and indeed it is lethal to be black (as in unpolluted woods). No melanic moth will survive. If $s = +1$, the melanic form has a huge advantage; indeed it is lethal to be normal (as in polluted woods). No normal moths will survive. If $s = 0$, neither form has an advantage over the other.

Note that both the *MM* and *Mn* moth numbers are reduced by the factor α. The next generation of moths, then, starts out with relative genotype frequencies

$$MM : Mn : nn = \alpha x^2 : 2\alpha x(1 - x) : \gamma(1 - x)^2 \tag{2}$$

or, to normalize these proportions so that their sum is 1

$$MM : Mn : nn = \frac{\alpha x^2}{\sum} : \frac{2\alpha x(1 - x)}{\sum} : \frac{\gamma(1 - x)^2}{\sum} \tag{3}$$

where

$$\sum = \alpha x^2 + 2\alpha x(1 - x) + \gamma(1 - x)^2 =$$
$$(\gamma - \alpha)x^2 - 2(\gamma - \alpha)x + \gamma \tag{4}$$

Each *MM* moth contributes two *M* genes to the gene pool, and each *Mn* moth contributes one *M* gene. Therefore, the fraction of *M* genes in the new generation is $\frac{\alpha x^2}{\sum} + \frac{\alpha x(1 - x)}{\sum}$, or $\frac{\alpha x}{\sum}$.

Now by inverting equation (1) we find that

$$\frac{\gamma}{\alpha} = \frac{1 - s}{1 + s} \tag{5}$$

By using this, we can now express the gene frequencies in the new generation in terms of the selective advantage. Recall that in the initial generation the relative gene frequency was

$$M : n = x : 1 - x \qquad (6)$$

In the new generation it is

$$M : n = \frac{(1+s)x}{1 - s + 4sx - 2sx^2} : \frac{1 - s + (3s - 1)x - 2sx^2}{1 - s + 4sx - 2sx^2} \qquad (7)$$

We can apply this to generation after generation to see how the proportion of M gene changes from generation in terms of the selective advantage (or disadvantage).

In Figure 21, I start with a fraction $x = 0.001$ of M genes, and I watch the growth of this fraction with generation number for ten positive values of selective advantage—i.e. advantage to melanic moths on soot-covered tree trunks. Even for a mild advantage ($s = 0.1$), the fraction of M genes soon grows, while for a large advantage ($s = 0.9$) the growth of the M gene fraction is very rapid indeed, and this is believed to have happened to the moth Biston betularia in industrial areas in Britain. Note, however, that even if $s = 1.0$ (all normal phenotypes discovered and eaten by birds), the n gene survives (albeit in small numbers) because it is hidden and protected in the heterozygous Mn moths, which are phenotypically melanic.

Figure 21: Complete dominance of M over n. Growth of the M—gene fraction x with generation number for ten selective advantages, from $s = 0.1$ to 1.0 in steps of 0.1.

What happens if we start with a high proportion of melanic genes, say $x = 0.999$, and put the moths in an unpolluted wood, where the tree trunks are lichen-covered, and the melanics are at a selective disadvantage (s is negative)? This in fact appears to be happening now in England, where air pollution controls are resulting in lichens recolonizing tree trunks that had become blackened with soot in a less environmentally

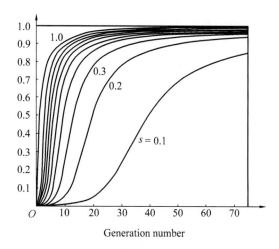

Figure 21　Generation number

— conscious era. Well, if we do the calculations, starting with $x = 0.999$, we find that almost nothing happens unless $s = -1$ exactly, in which case being a melanic phenotype is a death sentence, whether genotypically MM or Mn. The melanic gene is immediately extirpated. However, for any other negative value of selective advantage, very little happens for many generations, and the population remains predominantly melanic. This is because, even though the normal moths have the advantage, there are hardly any of them to enjoy it. Thus if the fraction of genes that are M is 0.999, the fraction of moths that are normal is only $(0.001)2$, or $0.000\,001$. For example, if we start with the fraction of M genes $x = 0.999$, and put them under a severe selective disadvantage of $s = -0.9$, even after 50 generations x is still $0.992\,7$. However, after x has dropped to about 0.95, and normal (advantaged) moths begin to appear in the population in appreciable numbers, the decline of the M gene is rapid or even catastrophic. (Is this why the dinosaurs suddenly vanished after a long period of world dominance? Just a thought!) Indeed, since the disadvantaged M gene is not hidden and protected in the heterozygous moth, the M gene is eventually completely extirpated. In Figure 22, I have started with $x = 0.9$ (which is low enough for the start of rapid decline after a long period of quasistability), and we follow the decline of the M gene for a further 75 generations for 10 negative values of selective advantage.

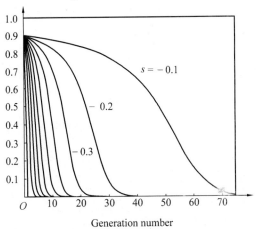

Figure 22 Generation number

Figure 22: Complete dominance of M over n. Decline of the M-gene fraction x with generation number for ten selective advantages, $s = -0.1$, to -1.0 in steps of 0.1.

So far, we have considered the case where one gene is completely dominant over another—but this is not always the case. In some species of moth the heterozygous form is intermediate in appearance to the two homozygous forms. In that case I'll use a small m for the "melanic" gene and a small n for the "normal" gene, so as not to give the impression that one is dominant over the other. The three possible genotypes are then mm, mn and nn, and they correspond to three phenotypes, melanic, intermediate and normal. I need to define selective advantage for each of the three forms, which I do as follows. I suppose that when one generation hatches from eggs, the relative numbers of genotypes in the population are in the proportion

$$mm : mn : nn = X : Y : Z \qquad (8)$$

Let us suppose that, by the time these moths lay their eggs to start the next generation, the numbers of melanic, intermediate and normal moths have been reduced by fractions α, β, and γ respectively. Then I define the selective advantages of the three forms as follows

$$mm : \quad s_1 = \frac{2\alpha - \beta - \gamma}{2\alpha + \beta + \gamma} \qquad (9)$$

$$mn : \quad s_2 = \frac{2\beta - \gamma - \alpha}{2\beta + \gamma + \alpha} \qquad (10)$$

$$nn : \quad s_3 = \frac{2\gamma - \alpha - \beta}{2\gamma + \alpha + \beta} \qquad (11)$$

These are not independent, and it takes a little algebra to show that they are related by

$$s_1 s_2 s_3 - 2(s_2 s_3 + s_3 s_1 + s_1 s_2) + 3(s_1 + s_2 + s_3) = 0 \qquad (12)$$

They all have the property that they are in the range -1 to $+1$. A value of $+1$ means that the other two genotypes are completely destroyed, whereas a value of -1 means that that genotype is completely destroyed.

We can then do just what we did before when we went from equation (1) to equation (3). We suppose that the gene ratio of one generation is $x : 1 - x$. Then it works out that the fraction of m genes in the next generation is

$$\frac{(\alpha - \beta) x^2 + \beta x}{(\alpha - 2\beta + \gamma) x^2 + 2(\beta - \gamma) x + \gamma} \qquad (13)$$

Readers might like to convince themselves why it is not possible to invert equations (9) – (11) to express α, β, and γ uniquely in terms of the selective advantages, which is why it is more convenient and informative to write equation (13) in terms of α, β, and γ. One can then easily get a computer to apply this formula through generation after generation and see how the fraction of m genes changes with generation number.

There are four qualitatively different cases to consider.

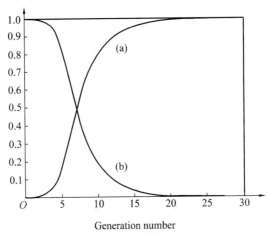

Figure 23   Generation number

Figure 23: No dominance. (a) Case I. mm is the fittest, nn is the least fit. (b) Case II. mm is the least fit, nn is the fittest.

(1) The homozygous melanic mm is the fittest, and the homozygous normal nn is least fit. That is $\alpha > \beta > \gamma$. In Figure 23(a) I illustrate this for $\alpha = 0.4$, $\beta = 0.3$, and $\gamma = 0.1$. These correspond to $s_1 = +0.33\overline{3}$, $s_2 = +0.0\overline{90}$, and $s_3 = -0.55\overline{5}$. I start with $x = 0.001$. The proportion of the m gene rapidly increases and the n gene eventually becomes extinct.

(2) The homozygous melanic mm is least fit, and the homozygous normal nn is fittest. That is $\alpha < \beta < \gamma$. In Figure 23(b) I illustrate this for $\alpha = 0.1$, $\beta = 0.3$, and $\gamma = 0.4$. These correspond to $s_1 = -0.55\overline{5}$, $s_2 = +0.0\overline{90}$, and $s_3 = +0.33\overline{3}$. I start with $x = 0.999$. The proportion of the m gene rapidly decreases and eventually becomes extinct.

Figure 24: The heterozygous form has the advantage. Regardless of

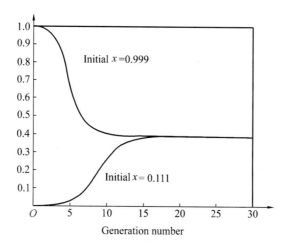

Figure 24   Generation number

the initial m-gene fraction, high or low, an equilibrium situation ultimately results.

(3) The heterozygous form has the advantage. That is $\beta > \alpha$ and $\beta > \gamma$. This case is rather more interesting! Regardless of the initial value of m-gene fraction x, the m-gene fraction eventually settles down to an equilibrium value x_e given by

$$x_e = \frac{\beta - a}{2\beta - \alpha - \gamma} \qquad (14)$$

If the m − gene fraction is initially higher than this, it drops to the equilibrium value; if it is initially lower than this, it rises to the equilibrium value. This presumably means that if you have a population in which the three forms exist together for a long time, the heterozygous form is fitter than the other two. This case is illustrated in Figure 24, which I calculated for $\alpha = 0.2$, $\beta = 0.8$, and $\gamma = 0.4$. These correspond to $s_1 = -0.500$, $s_2 = +0.4\overline{54}$, and $s_3 = -0.11\overline{1}$. I started with $x = 0.001$ and $x = 0.999$.

(4) The heterozygous form is at a disadvantage. That is $\beta < \alpha$ and $\beta < \gamma$. Can you guess what will happen, just by thinking about it without

actually doing the calculations? What happens is that there is still an equilibrium m-gene fraction, and it is still given by equation (14)—but it is an unstable equilibrium! If the m-gene fraction starts ever so slightly above this equilibrium value, the fraction grows until the n-gene becomes extinct; and if the m-gene fraction starts ever so slightly below the equilibrium value, the m-gene becomes extinct. This case is illustrated in Figure 25, which I calculated for $\alpha = 0.8$, $\beta = 0.2$, and $\gamma = 0.5$. These correspond to $s_1 = +0.391$, $s_2 = +0.529$, and $s_3 = 0.000$. I started with $x = 0.3333$ and $x = 0.3334$. Very slight differences in initial conditions result in quite different outcomes.

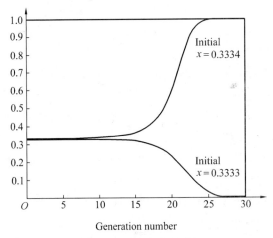

Figure 25 Generation number

Figure 25: No dominance. The heterozygous form is at a disadvantage. The m-gene fraction goes to zero or infinity depending on whether its initial value is below or above a critical value.

Of course we have so far looked at some highly idealized situations. For example, we have assumed that the selective pressures remain constant generation after generation. If the environment changes at some time, this poses no particular difficulty: we can change the values of the selective advantage at any generation and resume the calculation with these new values. There is one example of biological significance that is also particularly amenable to this sort of mathematical calculation, and that is the study of mimicry. Some butterflies taste nasty and they are brightly coloured (warning coloration) so that birds can easily recognize them and leave them alone. Most tasty insects are cryptically coloured—difficult to find. But there are a few cheats. Some tasty insects mimic the bright colours of their horrible-tasting cousins; birds see the bright colours of the mimics and assume that they taste awful, so they leave them alone. This mimicry gives the cheat quite a selective advantage. But the cheat is effective only if the mimic is much rarer than the model. If the cheats are abundant, birds will not be taken in so easily and will soon unmask the fraud.

We can construct a plausible mathematical model of this situation. Let us suppose that there is a gene M for mimicry and a gene n for non-mimicry. To keep things simple, we'll suppose that the gene M is dominant over n (as you already guessed from the capital and small letters), so that there are just two forms of the insect—a mimetic form, which can be either MM or Mn, and a non-mimetic form. At some time, the fraction of mimetic insects in the population is X and the fraction of non-mimetic insects is $1 - X$. The selective advantage, we suppose, depends on the value of X, as we argued in the previous paragraph. Suppose, for example, that

$$s = -\frac{1}{2} + (1 - X)^{\frac{1}{2}} \quad (15)$$

I have chosen this function quite arbitrarily, but it is at least plausible. It means that when the mimetic form is very rare (X very small), it has a distinct advantage ($s = +\frac{1}{2}$), and when it is common (X close to 1) it is at a decided disadvantage ($s = -\frac{1}{2}$). It has neither advantage nor disadvantage ($s = 0$) when $X = 0.75$. I admit that I also chose the function because it gives a very simple relation between s and x, the fraction of genes that are M. It is easy to show that

$$s = \frac{1}{2} - x \qquad (16)$$

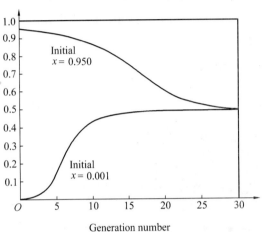

Figure 26 Generation number

Figure 26: Complete dominance. The selective advantage of the melanic form depends upon its relative abundance in the population.

Thus in terms of gene fraction (rather than mimetic insect fraction), s decreases linearly with x, going from $+\frac{1}{2}$ to $-\frac{1}{2}$, becoming zero for $x = \frac{1}{2}$. We can anticipate that, whatever the initial gene fraction, it will either increase or decrease until it reaches an equilibrium value of $+\frac{1}{2}$, when there is no selection but merely equal predation on mimetic and nonmimetic forms. The calculation is very easy. We just use equation (7) as before, but, instead of a constant value of s, we substitute $\frac{1}{2} - x$. The behaviour is illustrated in Figure 26, for initial gene fractions of

0.950 and 0.001.

If you wish, you can further elaborate on these models of evolution by variation and natural selection and watch the changes in the population of the moths before your very eyes. I suppose it goes to show that, whatever subject happens to interest you, you will probably always find some application of mathematics to it that will make it even more interesting.

Words and Expressions

moth *n*. 虫,蛾

eccentric *adj*. 古怪的,怪癖的;不同中心的(圆);离中心的,偏心的

caterpillar *n*. 蝴蝶或飞蛾类之幼虫;毛虫

ultimately *adv*. 最后;最终;终结

mathematical analysis 数学分析

disprove *vt*. 反证,证明为误

lichen *n*. 青苔,地衣,苔藓

cryptic *adj*. 秘密的,隐藏的,潜隐的

coloration *n*. 染色;着色;色泽

cryptic coloration 保护色

melanic *adj*. 黑变病的,黑性的,黑色素过多的

conspicuous *adj*. 显著的,显而易见的,引人注目的

grave *n*. 坟墓,墓碑;死地,死 *adj*. 庄重的,严肃的,重大的;阴沉的

belch *vt*. 吐气,吐出 *vi*. 打嗝,喷出 *n*. 大嗝,嗝;喷出之物

grime *n*. 污秽 *vt*. 使污浊,使覆有污秽物

genetics *n*. 遗传学

geneticist *n*. 遗传学者

inherit *vt*. 继承,由遗传而得 *vi*. 继承,接受遗传的力量、特点等

genotype *n*. 基因型,遗传型

homozygous *adj*. 同型结合的,纯合子的

intermediate *adj*. 中间的,中级的,中间人的 *n*. 中间物,中间人,调停者

dominant *vt.* 统治,支配 *vi.* 管辖,支配
recessive *adj.* 隐性的,退后的
bog *n.* 沼泽 *vt.* 使陷于泥潭
vulnerable *adj.* 可伤害的,易受攻击的;难防守的,有弱点的
snap *vt.* 咬,攫取
predominantly *adv.* 主要地,杰出地
phenotypically *adv.* 显型地
quantitatively *adv.* 与量有关地,定量地
factor *n.* 因数,因子,遗传因子
lethal *adj.* 致命的,致死的,毁灭性的
invert *vt.* 上下颠倒,倒转,转回
frequency *n.* 频率,时常发生
soot *n.* 煤烟,煤灰,油烟 *vt.* 使某物蒙上黑烟灰
mild *adj.* 温柔的,和善的,温暖的,宽大的,不严重的
albeit *conj.* 虽然
phenotype *n.* 显型,表型;显型类之所有生物
recolonize *vt.* 再殖民于,重新殖民于
conscious *adj.* 觉得的,知道的;有意识的
extirpate *vt.* 根除,灭绝,连根拔起
appreciable *adj.* 可见到的,可觉得的,可估计的
catastrophic *adj.* 悲惨的,毁灭的
dominance *n.* 统治,支配
quasistability *n.* 准稳定的,类似稳定的
informative *adj.* 有益的,供给知识的,
extinct *adj.* 灭种的,灭绝的;熄灭了的
equilibrium *n.* 平均,均衡,均势
equilibrium value 平衡值
mimicry *n.* 拟态,模仿
cryptically *adv.* 隐性地,隐蔽地,秘密地
mimic *vt.* 仿效,拟态;与…极相似 *n.* 擅模仿的人或物

6.3 RECKONING AND REASONING

You might have heard of this story, but it bears being repeated. In 1992, Lou D'Amore, a science teacher in the Toronto area, sprung a Grade 3 arithmetic test from 1932 on his Grade 9 class, and found that only 25% of his students could do all of the following questions.

1. Subtract these numbers: 9 864 − 5 947
2. Multiply: 92 × 34
3. Add the following: 126.30 + 265.12 + 196.40
4. An airplane travels 360 kilometers in three hours. How far does it go in one hour?
5. If a pie is cut into sixths, how many pieces would there be?
6. William bought six oranges at 5 cents each and had 15 cents left over. How much had he at first?
7. Jane had $2.75. Mary had 95 cents more than Jane. How much did Jane and Mary have together?
8. A boy bought a bicycle for $21.50. He sold it for $23.75. Did he gain or lose and by how much?
9. Mary's mother bought a hat for $2.85. What was her change from $5?
10. There are 36 children in one room and 33 in the other room in Tom's school. How much will it cost to buy a crayon at 7 cents each for each child?

This modest quiz quickly rose to fame as "The D'Amore Test." Other teachers tried it on their classes, with similar results. There was some improvement in Grades 10 to 12, where 27% of students could get through it, but they tend to be keener anyway since their less ambitious class-mates usually give up on quantitative science after Grade 9. All in all, the chance of acing the D'Amore Test appears to be independent of anything learned in high school.

At first glance this seems as it should be, because the test certainly contains no "high school material". On second thought, however, a strange asymmetry appears: while all students expect to use the first two R's (Readin' and Ritin') throughout their schooling and beyond, they drop the third R (Rithmetic) as soon as they can—if indeed they acquired it at all. Has it always been like this? I doubt it: my grandmother went to school only twice a week (being needed in yard and kitchen) but was later able to handle all the arithmetic in her little grocery store without prior attendance of remedial classes. She did not even have a cash register.

To many administrators, think-tankers, etc., this is beside the point, because we now live in the brave new computer age. A highly placed person who has likely never repaired a car engine, and probably knows little about computers, suggested that 20 years ago, "an auto mechanic needed to be good at working with his hands", whereas now he needs Algebra 11 and 12 to run his array of robots.

Mechanics laugh at this: remember the breaker-point gaps, ignition timing, engine compression, battery charge, alternator voltage, headlight angle, and a multitude of other numerical values we had to juggle in our minds and check with fairly simple tools — today's gadgets make our jobs more routine, they say. But ministerial bureaucrats tend to believe the hype, with a fervour proportional to their distance from "Mathematics 12", which has gobbled up Algebra 12 in most places.

Aye, there's the rub: the third R has morphed into the notorious M. "What's in a name?" you ask, "that which we called rithmetic by any other word would sound as meek." How many times must you be told that M is hard and boring, and hear the refrain "I have never been good at M"? It is the perfect cop-out, acceptable even in the most exclusive company — a kind of egalitarian salute by which "normal" members of the species homo sapiens recognize one another. How can a teacher of, say, social studies be expected to develop vivid lessons around unemployment, national debt, or global warming—as long as these topics are mired in M? He/she still must mention numbers, to be sure, but can now present them in good conscience as disconnected facts, knowing that his/her students' minds will be uplifted in another class, by that lofty but (to him/her) impenetrable M.

Ask any marketing expert: labels are not value-free, they attract, repel, or leave you indifferent. Above all, they raise expectations, which, in the case of M, are as manifold and varied as the subject itself. Is it conceptualization, exploration, visualization, constructivism, higher-order thinking, problem solving or all of the above? The guessing and experimenting goes on and on, producing bumper crops of learned papers and theses, conferences, surveys, and committees, as well as confused students and teachers. "This is the first time in history that Jewish children cannot learn arithmetic" said an Israeli colleague, referring to the state of Western style education in his country, where the recent Russian immigrants maintain a parallel school system.

Not every country has followed the R to M conversion. In the Netherlands and (what was) Yugoslavia, children still learn rekenen and račun, respectively, together with reading and writing. The more weighty M is left for later. Germany clung to Rechnen till the 1960's, and then rashly followed the American lead, pushing Mathematic all the way down to Kindergarten-with the effect of finding itself cheek-to-jowl with the US in international comparisons.

I hear the sound of daggers being honed: what is this guy trying to sell (in this culture we are all vendors), is it "Back to Basics"? Does he hanker for "Drill and Kill," for "Top Down" at a time when all good men and women aspire to "Bottom Up"? Readers unaccustomed to Educators' discourse might be puzzled at such extreme positions getting serious attention. They would immediately see middle ground between tyranny and anarchy, boot camp and nature trail, etc. Why do we always argue Black versus White? I really cannot explain it. Maybe it is because we need strident voices and must hold single notes as long as we can, in order to be noticed in this mighty chorus. How did we get here?

Although the benefits of planned obsolescence are obvious, they are not often mentioned to justify the present trend toward innumeracy. It is the relentless advance of technology which must be seen as the main reason for the retreat of archaic skills. Speech-recognizing computers already exist, and once they are mass-produced, writing will not need to be taught anymore, at least not at public expense. Whatever we now do with our hands and various other body-parts outside the brain will clearly fall into the domain of sports. Only in this spirit does it make sense to climb a mountain top that can be more safely reached by helicopter.

Before the advent of electric and later electronic calculators, computations had to follow rigid algorithms that allowed the boss or auditor to check them. This was "procedural knowledge" of an almost military kind—justly despised and rejected when it became obsolete. Oddly enough it did, however, have an important by-product: by sheer habit, simple calculations were done at lightning speed, and often mentally—of course with a large subconscious component. In many places, this "mental arithmetic" was even practised as a kind of sport, still "procedural," in some sense, but open to improvisation — more like soccer than like target shooting.

Look at the first question of the D'Amore Test: 9 864 – 5 947. Abe did it the conventional way and had to "borrow" twice. Beth zeroed in on the last three digits, noting that 947 exceeded 864 by 36 + 47 = 83, which she subtracted from 4 000. Chris topped up the second number by 53 to 6 000 and hence had to increase the first one to 9 864 + 53 = 9 917. Dan and Edith had yet different ways, but all got 3 917. On the second question, Abe again used the standard method, since he was a bit lazy but meticulous. Beth looked at the 92 and thought $100 - 10 + 2$, playing it very safe. Chris spotted one of his favourite short-cuts: $3 \times 17 = 51$, and reasoned that $9 \times 34 = 6 \times 51 = 306$, and so on. Dan was attracted to the fact that 92 was twice 46, which lies as far above 40 as 34 lies below it. Therefore 46×34 was $1\,600 - 36$, which had to be doubled to $3\,200 - 72$. Edith blurted out the answer $3\,128$ and said she did not remember how she got it.

When I was in Grade 7, I knew such kids—and was irked by the fact that many played this mental game as well as they played soccer. Justice was restored when, in Grade 8, they were left in the dust by x and y but continued to outrun me on the playing field. Maybe they never missed the x and y in later life (unlike contemporary plumbers), but I am almost sure their "number sense" often came in handy. Today's kids are to acquire this virtue by doing brain-teasers and learning to "think like mathematicians," carefully avoiding "mindless rote."

Whenever I walk by the open door of a mathematician's work place, I see black or white boards covered with calculations and diagrams. How come they get to indulge in this "rote," while kids must fiddle with manipulations or puzzle till their heads ache? Could it be that we mathematicians sometimes engage in "mindful rote"—the kind known to musicians and athletes? If so, we ought to step out of the closet and tell the world about the joy of rote. Anyone who has observed young children will immediately know what we mean.

And while we're at it, we might reclaim ownership of the M-word, at least suggest that it be kept out of the K-4 world. This does not mean that schools should go back to teaching arithmetic—admittedly an awkward label. How about "reckoning and reasoning," a third and fourth R to balance the first two? They would be associated with good old common sense, and, as Descartes has pointed out, nobody ever complains of not having enough of that.

Words and Expressions

reckon *vt*. 计算,估算 *vi*. 计算,算账
reckoning *n*. 计算,统计,核算
reason *n*. 原因,前提;理由,推理 *vt*. 说服,劝说
reasoning *n*. 推理,推论,推导,推断
quiz *n*. [美]非正式的考试
keen *adj*. 热心的,渴望的;锋利的,尖锐的
keener *n*. 热心者
quantitative *adj*. 数量的,关于数量的
asymmetry *n*. 非对称
remedial *adj*. 修补的,矫正的;补救的;治疗的
administrator *n*. 管理者,治理者,行政官
array *n*. 排列;整列,排阵;序列 *vt*. 排列,布置;装饰
ignition *n*. 点火,燃烧,点火装置
alternator *n*. 交流发电机
voltage *n*. 伏特数,电压,电压量
multitude *n*. 众多,群众
gadget *n*. [俗]设计精巧的小机器
routine *n*. 例行公事,惯例,常规
ministerial *adj*. 牧师、内阁或部长的,部长级的;行政上的;附属的
bureaucrat *n*. 官僚
fervour *n*. 热诚,热心;热,炙热
gobble *vt. or vi*. 大吃,狼吞虎咽
notorious *adj*. 声名狼藉的,著名的,众人皆知的
egalitarian *adj*. 相信人人平等的 *n*. 平等主义
salute *n*. 致意,问候;敬礼 *vt*. 向…敬礼,向…致意 *vi*. 致意,祝贺
mire *n*. 泥泞,沼池,泥浆,困境 *vt*. 使陷入泥泞,使陷入困境
lofty *adj*. 高的,高尚的;高傲的
impenetrable *adj*. 不可理解的,难以探究的

manifold *adj*. 多重的,繁多的;多倍的,同时做多种事 *n*. 簇,流形;复写本;复写纸;多重

conceptualization *n*. 概念化

exploration *n*. 探险,仔细检查

visualization *n*. 想像

bumper *n*. 满杯; *vt*. 倒满杯

Israeli *n*. 以色列;犹太人

Netherlands *n*. 荷兰

Yugoslavia *n*. 南斯拉夫

hone *vi*. 发怨声,悲叹 *n*. 细磨刀石

hanker *vi*. 渴望,眷恋

tyranny *n*. 虐政;暴行

anarchy *n*. 无政府状态,混乱

versus *prep*. 对;相形,比较

strident *adj*. 作粗糙声的,发尖锐声的

obsolescence *n*. 将要过时

innumeracy *n*. 无数学基本功,不识数

relent *vi*. 变温和,变宽容 *vt*. 使变温和,使变宽容

relentless *adj*. 无慈悲心的,不放松的

archaic *adj*. 古的,古代的;已废的,已不通用的

advent *n*. 到来,来临

rigid *adj*. 僵硬的,稳固的;严格的,严厉的;精确的

algorithm *n*. 算法,计算程序

procedural *adj*. 程序的,程序上的,有关程序的

despise *vt*. 轻视,蔑视

obsolete *adj*. 作废的,已废的;过时的

subconscious *adj*. 潜意识的 *n*. 潜意识,下意识

component *adj*. 组成的,成分的 *n*. 成分,分力

mental arithmetic 心算

improvisation *n*. 即席创作;即兴

meticulous *adj*. 极端注意琐事的,拘泥于细节的
blurt *vt*. 不假思索地脱口说出 *n*. 不假思索的话
indulge *vt*. 放任,纵容;迁就 *vi*. 纵情,任意
rote *n*. 机械的方法,固定程序;背诵,强记
fiddle *vi*. 做无益之事,乱动 *vt*. 演奏,虚度(光阴)
manipulation *n*. (用手的)操纵;巧妙的操纵

6.4 AM I REALLY SICK?

When Nelly came back from her year in Ladorada, she read in The New York Times that tuberculosis was on the rise again, especially in the part of the world she had just visited. Her doctor explained to her that there was not much to worry about, as only 0.01 percent of the inhabitants of that beautiful and hospitable land were affected, but that it was wise to take a simple test, which he had right there in his office. The next day, he phoned her to say that, unfortunately, she had tested positive. "Does that mean I am really sick?" she asked. "It doesn't look good," said the doctor. "The test is 99.9 percent sure." Her brother Nick, who was an engineer with a good feel for numbers, asked her to quote those figures again, and then burst out laughing: "only one in 10 000 Ladoradans is infected, and that dumb test makes one mistake in 1 000 trials, see?" Nelly shook her head: "No, I only see a guy who giggles at numbers. Tell me what's so funny." Nick gave her that big brother look: "According to your doctor, the test is right 999 times in 1 000, on average. So, if you test 10 000 people from over there, you'll get 10 false alarms and one real case—on average." That evening, Nelly took a long walk in Central Park mulling over Nick's argument that her chances of being sick were only one in 11. What a relief!

She had dinner with her friend Cornelia, who was studying to be a nurse. "I can't get over it," said Cornelia. "In class today we just began to study that new TB outbreak in Ladorada as well as the Kinski Test, and here I am having dinner with a real specimen — that's so cool." Nelly recalled seeing the name Kinski on the box in Dr. Dixit's office. "It's a new test," Cornelia rambled on, "99.99 percent accurate. But don't worry, Nelly: even if you test positive, the chances you are really infected are only 50 percent. That's what our instructor said — always 50 percent — and he does research in a big lab." Nelly was aching to run Nick's logic through her mind. "My doctor said its accuracy was only 99.9 percent," she ventured. "So what?" Cornelia shot back, "In either case, we have near certainty. Remember, you always have a 50-50 chance to be a false positive." Nelly remarked that Nick had convinced her that the chance was 10 in 11, if Kinski was only 99.9 percent certain.

Though she had always had an eye on Nelly's brother, Cornelia now snapped that he was just an electrical engineer. "What does he know about epidemiology?" She spoke that word with the solemnity of a neophyte. To save the evening, the two women opened another bottle of wine.

As she was waking up the next morning, Nelly was able to reconstruct Nick's argument for a test that was 99.99 percent accurate. It would make one false diagnosis in 10 000 cases, and in Ladorada there would be one real TB-carrier among them—on average, as Nick would say. Hence you would likely find two "positives" in that crowd. The reasoning behind Cornelia's "50-50" chance was therefore the equality of the infection rate in the country and the failure rate of the test, namely one in 10 000. Good! No, bad!

It meant that she had better do something: 50 percent was too close for comfort. She got an appointment with an X-ray lab, but only after the impending long weekend, most of which she spent practicing her violin. After all, TB was not AIDS, although the new strain was said to be particularly resistant to treatment. But time and again, she was drawn to search in the Internet for news on TB in Ladorada.

On Monday, still a holiday, she found a reputable Spanish site describing the uneven spread of tuberculosis in Ladorada: the capital was stricken 10 times harder than the country as a whole. "One in 1 000 for the likes of me," she thought, because she had spent almost all her time in Hermosa. She reviewed Nick's reasoning: if 10 000 inhabitants of that city were tested, 10 true positives would turn up — on average — because of the infection rate, and one false positive because of the margin of error in Kinski. The tables had turned: her chances of being healthy were down to one in 11. What a bummer! Just then Cornelia phoned to say how sorry she was to have been so snarky about Nick. Nelly told her about the new odds.

"Don't worry so much," Cornelia suggested. "The experts say the odds are 50-50 for false positives; that's what you should go by, instead of confusing yourself with simplistic calculations." The conversation ended with some chatter about Nick's vulnerability to predatory females. Quite a pair of health professionals, Dr. Dixit and Cornelia, thought Nelly. But the word "simplistic" struck a chord. What about the false negatives—sick people given a clean bill of health by the test. Wouldn't they diminish the 10 "true" positives?

Not by much, of course, but Nelly was suddenly more interested in the calculation than in her own health. She phoned her friend Fatima, a graduate student in statistics, and was glad to find her at home. After hearing what the problem was, Fatima invited her over for tea. "It's a classic," she smiled, "I have to explain it in my tutorials every single year, so I made this slide to put on the overhead." Nelly vaguely made out some letters, lines, and dots. "We'll go over it after tea," said Fatima.

"See this dot labelled TB here? It represents the infected part of the population, and the letter r stands for the odds of being infected, $\frac{1}{1\,000}$ in your case, but it could be $\frac{38}{31\,570}$ or something crazy like that: it's best to think of it as a percentage. And the $1 - r$ is the opposite percentage, the chance of being uninfected, hence it's connected to the dot labelled OK." All that was straightforward, but didn't tell you anything, thought Nelly. "How come the dot is labelled TB?" she asked. "Oh, just because it has only two letters. The so-called infection could be any hidden condition you want to ferret out with your test-like a secret yearning," Fatima whispered, "Say, for chocolate. But let's get back to the test."

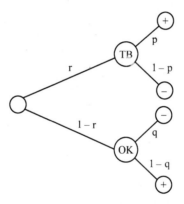

Figure 27

"The test marks you either as positive or negative." Now Nelly began to catch on. "And there are two of each since you could be a true or a false positive or negative. Could I try to explain the rest of the diagram?" Fatima was delighted to have such an eager student. "If you are a TB, the chances of being correctly identified are p; if you are a OK, they are q. Why are they not the same?" Fatima pointed out that the true TB_s were usually easier to identify than the true OKs, so p was usually bigger than q. "But in the Kinski Test, they are the same: 99.99

percent, aren't they?" said Nelly. "I looked it up after you phoned me," replied Fatima, "p is indeed 99.99 percent but q only 99.9 percent—so the test typically produces one positive out of every 1 000 OK_s." Nelly threw her arms around Fatima: "That means I'm back to 50 percent, if Nick's reasoning is correct. Oh, Fatima, please tell me that it is!" Fatima thoroughly enjoyed being the bringer of glad tidings, but said: "Not quite. Let's work it out all the way."

They tallied the positives: the true ones from TB were $r \times p$ and the false ones from OK were $(1-r) \times (1-q)$, for a total of $(1-r)(1-q) + rp$. Thus, the chances of a false positive were 1 in $\frac{1+rp}{(1-p)(1-q)} = 1 + \frac{r}{(1-q)}\frac{p}{(1-r)}$. "Nick neglected that last factor $\frac{p}{(1-r)}$," said Fatima, "But look at it: $\frac{99.99}{99.9}$ in this case. It isn't much of a factor, and that is typical: for any half-decent test it's very close to one." Nelly had tears in her eyes and didn't know whether they came from the 50 percent or from understanding the calculation. "Fatima, you are an angel," she said, "But tomorrow I'll still have myself checked out." Back on the street, the only thing that irked her was that the news would strengthen Cornelia's blind faith.

Words and Expressions

tuberculosis $n.$ ($abbr.$ TB)结核病,(尤指)肺结核
inhabitant $n.$ (某地的)居民,住户,栖息的动物
ramble $v.$ 漫步,闲逛;漫谈,闲聊
ventured $n.$ 工作项目或事业,商业,企业 $v.$ 敢于去,敢说
solemnity $n.$ 庄严,严肃
neophyte $n.$ 初学者
diagnosis $n.$ 诊断,判断问题,确定毛病;诊断结果,诊断书
infection rate 感染率

impending *adj*. 即将发生的,迫在眉睫的,行将到来的
reputable *adj*. 声誉好的,有名望的,值得信赖的
bummer *n*. [美俚]依赖他人而过活的人,懒虫,饭桶
simplistic *adj*. 过分简单的
vulnerability *n*. 弱点,容易受伤害
predatory *adj*. 抢夺的,有抢夺性的
diminish *vt*. 减少,缩小,使变小
statistics *n*. 统计学;统计,统计表
tutorial *adj*. 家庭教师的,(大学)导师的
slide *vi*. 滑动;滑行,溜过 *vt*. 使滑动,使轻轻溜过
vaguely *adv*. 不明确地,不清楚地;无表情地
yearn *vi*. 渴望,思念;怀念,怜悯
decent *adj*. 适当的,可接受的;令人满意的,得体的

7

Vector Space
Some Problems Over Skewfield

7.1 BASIC CONCEPT

We always use K to denote skewfield of degree n, $K^* = K \setminus \{0\}$, $Z = \{0, \pm 1, \pm 2, \cdots\}$, $Z = \{1, 2, \cdots\}$, $m, n \in Z^+$, we use $K_{m,n}$ to denote the set of matrix of (m, n)-type, $K_n = K_{n,n}$, $I_n \in K_n$ is unit matrix. Let $A \in K_n$, If exist $B \in K_n$, such that

$$AB = BA = I_n$$

then we call A is a invertible matrix of degree n over K, $GL_n(K)$ is the set of all invertible matrices of degree n, it is called a group under the common matrix multiplication, it is called general linear group of degree n over K.

Let $GL_n(K) \cong GL(V)$ in which V is n-dimension vector space over K. $n \geq 2$ we use $E_{i,j}$ to denote a matrix of degree n over K, whose (i, j) entry is 1, the others are 0, $i, j = 1, \cdots, n$, if $\lambda \in K$, let

$$T_{i,j}(\lambda) = I_n + \lambda E_{i,j}$$

$$i, j = 1, \cdots, n, \ i \neq j$$

when $\lambda \neq 0$, we call $T_{i,j}(\lambda)$ is a simple transvection of degree n over

K. Obvious $T_{i,j}(\lambda) \in GL_n(K)$ and $T_{i,j}^{-1}(\lambda) = T_{i,j}(-\lambda)$ as well as $n \geqslant 3$

$$T_{i,j}(\lambda) = [T_{i,k}(\lambda), T_{k,j}(1)]$$
$$k \in \{1, \cdots, n\}, k \neq i, j$$

$SL_n(K)$ who is a group generated from all the simple transvections of degree n over K is called special linear group over K.

A matrix A is a transvection of degree n if it is similar to a simple transvection in $GL_n(K)$. Thus the inverse and the conjugate element of a transvection is a transvection for all $P \in GL_n(K)$, in which A is a transvection. Get $A, B \in GL_n(K)$, denote $[A, B] = ABA^{-1}B^{-1}$. We call $[A, B]$ is a commutator generating by A, B. When A, B are all transvections, we call $[A, B]$ is a commutator of transvection. Easy to know, when $[A, B]$ is a commutator of transvection, $[A, B]^{-1}$, $P[A, B]P^{-1}$ are all commutators of transvection $\forall\ P \in GL_n(K)$.

Let $A \in GL_n(K)$, denote

$$\mathrm{res}A = \mathrm{rank}(A - I_n)$$

We called it residue number of A, clearly

$$\mathrm{res}A = 0 \Longleftrightarrow A = I_n$$

Proposition 1 Let $A_1, \cdots, A_m \in GL_n(K)$, $\forall\ m \in Z^+$, then
$$\mathrm{res}(A_1 \cdots A_m) \leqslant \mathrm{res}A_1 + \cdots + \mathrm{res}A_m$$

Proof We can from

$$A_1 \cdots A_m - I_n =$$
$$A_1 \cdots A_m - A_2 \cdots A_m + A_2 \cdots A_m$$
$$- A_3 \cdots A_m + \cdots + A_{m-1}A_m - A_m + A_m - I_n$$

through

$$\mathrm{rank}(A + B) \leqslant \mathrm{rank}\ A + \mathrm{rank}\ B, \forall\ A, B \in K_n$$

we can get the result. From the definition we have:

Proposition 2 Let $A \in GL_n(K)$, then
$$\mathrm{res}A^{-1} = \mathrm{res}A, \mathrm{res}(PAP^{-1}) = \mathrm{res}A. \quad \forall\ P \in GL_n(K)$$

for $A \in GL_n(K)$. The linear homogenous equations over K
$$(A - I_n)X = 0$$
it's solution space is called the fix space of A, denote P_A. Obviously, P_A is the subspace of the right vector space K^n over K and we also denote
$$R_A = \{(A - I_n)X \mid X \in K^n\}$$
R_A is a subspace of K^n, R_A is called the residue space of A. Obviously
$$\mathrm{res}A = \dim R_A$$
and easy to have
$$\mathrm{res}A = n - \dim P_A \tag{1}$$

Proposition 3 Let $A \in GL_n(K)$, $\mathrm{res}A = m > 0$, then exist P, $Q \in GL_n(K)$, such that
$$PAP^{-1} = \begin{pmatrix} I_{n-m} & B_1 \\ 0 & A_1 \end{pmatrix} \tag{2}$$
$$QAQ^{-1} = \begin{pmatrix} A_2 & 0 \\ B_2 & I_{n-m} \end{pmatrix} \tag{3}$$
in which
$$\mathrm{rank}\begin{pmatrix} B_1 \\ A_1 - I_m \end{pmatrix} = \mathrm{rank}\begin{pmatrix} A_2 - I_m \\ B_2 \end{pmatrix} \tag{4}$$

Proof From (1) $\dim P_A = n - m$, we get $X_1, \cdots, X_{n-m}$ are a set of basis of P_A through basis extension, we have $X_{n-m+1}, \cdots, X_n$ such that $X_1, \cdots, X_n$ are a set of basis of K^n. Let
$$P_1 = (X_1, \cdots, X_n)$$
then $P_1 \in GL_n(K)$, and
$$P_1^{-1}AP_1 = P_1^{-1}(X_1, \cdots, X_{n-m}, AX_{n-m+1}, \cdots, AX_n) =$$
$$\begin{pmatrix} I_{n-m} & B_1 \\ 0 & A_1 \end{pmatrix}$$
is the result. We can use the similar way to have the result of (3). (4) is obvious. The end.

Definition Let $A \in GL_n(K)$, when $\text{res}A > 0$, we call (2) (or (3)) is a res-form of A, in which A_1(or A_2) is called a residue matrix of A, I_n is the res-form of I_n.

Proposition 4 Let $A \in GL_n(K)$, $\text{res}A > 0$, then the arbitrary two residue matrix of A are similar to each other.

Proof When $\text{res}A = n$, the proposition is correct.

Next, we let $0 < m < n$, if the left of (2) and

$$\begin{pmatrix} I_{n-m} & B_2 \\ 0 & A_2 \end{pmatrix}$$

are all $A's$ res-form, then there exists $T \in GL_n(K)$, such that

$$T \begin{pmatrix} I_{n-m} & B_1 \\ 0 & A_1 \end{pmatrix} T^{-1} = \begin{pmatrix} I_{n-m} & B_2 \\ 0 & A_2 \end{pmatrix}$$

order

$$T = \begin{pmatrix} T_1 & T_2 \\ T_3 & T_4 \end{pmatrix}, \quad T_1 \in K_{n-m}$$

then

$$\begin{pmatrix} T_1 & T_2 \\ T_3 & T_4 \end{pmatrix} \begin{pmatrix} I_{n-m} & B_1 \\ 0 & A_1 \end{pmatrix} = \begin{pmatrix} I_{n-m} & B_2 \\ 0 & A_2 \end{pmatrix} \begin{pmatrix} T_1 & T_2 \\ T_3 & T_4 \end{pmatrix} \quad (5)$$

through direct calculate, we have

$$B_2 T_3 = 0, \quad (A_2 - I) T_3 = 0$$

since

$$\begin{pmatrix} B_2 \\ A_2 - I_m \end{pmatrix} T_3 = 0$$

through (4) we have

$$\text{rank} \begin{pmatrix} B_2 \\ A_2 - I_m \end{pmatrix} = m$$

hence $T_3 = 0$, $T_4 \in GL_n(K)$ through $(5) \Rightarrow T_4 A_1 T_4^{-1} = A_2$. The end.

We use $X \oplus Y$ to denote quasi-diagonal matrix $\begin{pmatrix} X & 0 \\ 0 & Y \end{pmatrix}$.

Proposition 5 Let $A \in GL_n(K), n \geq 2$, then A has a res-form $I_{n-m} \oplus A_1$, in which $m = \mathrm{res}A.$, if and only if the residue number of the residue matrix of A is $\mathrm{res}A$.

Proof Only to prove sufficiency.

Let (2) is correct, from
$$\mathrm{res}A_1 = m \Longrightarrow A_1 - I_m \in GL_m(K)$$
order $X = B_1(I_m - A_1)^{-1}$, then
$$\begin{pmatrix} I_{n-m} & X \\ 0 & I_m \end{pmatrix} \begin{pmatrix} I_{n-m} & B_1 \\ 0 & A_1 \end{pmatrix} \begin{pmatrix} I_{n-m} & X \\ 0 & I_m \end{pmatrix}^{-1} = \begin{pmatrix} I_{n-m} & 0 \\ 0 & A_1 \end{pmatrix}$$

The end.

We use $C(K^*)$ and $C(K)$ to denote the centres of K^* and K respectively, we use $C(SL_n(K))$ and $C(GL_n(K))$ to denote the centres of $SL_n(K)$ and $GL_n(K)$ respectively. Easy to know:

Proposition 6 $C(GL_n(K)) = \{cI_n \mid c \in C(K^*)\}$
$$C(SL_n(K)) = \{cI_n \mid c \in C(K^*), c^n \in D(K^*)\}$$
in which $D(K^*)$ is a commutator group of K^*.

Proposition 7 Let $A \in GL_n(K)$ then $A \bar{\in} C(GL_n(K))$ if and only if $P \in GL_n(K)$, such that
$$PAP^{-1} = \begin{pmatrix} 0 \\ 1 \\ 0 \\ \vdots & * \\ 0 \end{pmatrix}$$

Proof Only to prove the essential. From $A \bar{\in} C(GL_n(K))$, Suppose $a_{21} \neq 0$, let
$$a_1 = -a_{11}a_{21}^{-1}, \quad a_2 = a_{21}^{-1}, \quad a_j = -a_{j1}a_{21}^{-1}, \quad j \geq 3$$

$$P_1 = \begin{pmatrix} 1 & a_1 & & & \\ & a_2 & & & \\ & a_3 & 1 & & \\ & \vdots & & \ddots & \\ & a_n & & \cdots & 1 \end{pmatrix}$$

then $P_1 \in GL_n(K)$ and

$$PAP_1^{-1} = \begin{pmatrix} 0 & & \\ 1 & & \\ 0 & & \\ \vdots & & * \\ 0 & & \end{pmatrix}$$

if $a_{21} = 0$, but we have $i \in \{3, \cdots, n\}$ such that $a_{i1} \neq 0$. P_{st} is a matrix which is formed by altering I_n's sth and tth row (or column), we call it a simple matrix, then $(2, 1)$ entry of $P_{2i}AP_{2i}^{-1}$ is a_{i1}, form the above discussing $\Rightarrow$ the proposition is correct.

If $\exists i, j \in \{1, \cdots, n\}, i \neq j$, such that $a_{ij} \neq 0$, then there must have a nondiagonal element in the 1st column of $P_{1j}AP_{1j}^{-1}$ is not zero, so we can induce this to above discuss.

If $A = aI_n$, from $A \in C(GL_n(K)) \Rightarrow a \in (K^*)$. We have $b \in K^*$, such that $ab \neq ba$. So $(2, 1)$ entry of $T_{21}(b)AT_{21}^{-1}(b)$ is $ba - ab \neq 0$ the result is truth. The end.

Words and Expressions

group n. 群
linear group 线性群
skewfield n. 体
invertible adj. 可逆的
invertible matrix 可逆矩阵
general linear group 一般线性群

vector space 向量空间
entry n. 表值；进入，通路
transvection n. 平延
simple transvection 初等平延
special linear group 特殊线性群
inverse element 逆元
conjugate adj. 共轭的
conjugate element 共轭元
commutator n. 换位子
commutator of transvection 平延换位子
residue n. 剩余，留数
residue number 剩余数
homogenous n. 齐次的，齐的；同类的
linear homogenous equation 线性齐次方程
solution space 解空间
fix space 固定空间
right vection space 右向量空间
residue space 剩余空间
basis n. 基
extension n. 扩张，开拓；外延
residue matrix 剩余矩阵
diagonal adj. 对角的，对角线的
quasi-diagonal 拟对角的
quasi-diagonal matrix 拟对角矩阵
commutator group 换位子群

7.2 PROPERTY OF TRANSVECTION

  Theorem Let $A \in GL_n(K)$, $n \geqslant 2$, then the propositions below are equivalent

  1) A is transvection;

2) $\text{res}A = 1$, $A - I_n$ is nilpotent matrix;

3) For all $i \in \{1, \cdots, n\}$, exist $P \in GL_n(K)$ such that

$$PAP^{-1} = \begin{pmatrix} 1 & & & & & & \\ & \ddots & & & & & \\ & & 1 & & & & \\ \lambda_1 & \cdots & \lambda_{i-1} & 1 & \lambda_{i+1} & \cdots & \lambda_n \\ & & & & 1 & & \\ & & & & & \ddots & \\ & & & & & & 1 \end{pmatrix}$$

in which not all $\lambda_1, \cdots, \lambda_{i-1}, \lambda_{i+1}, \cdots, \lambda_n$ are zero.

4) For all $j \in \{1, \cdots, n\}$, exist $Q \in GL_n(K)$, such that

$$QAQ^{-1} = \begin{pmatrix} 1 & \cdots & & u_1 & & & \\ & \ddots & & \vdots & & & \\ & & 1 & u_{j-1} & & & \\ & & & 1 & & & \\ & & & u_{j+1} & 1 & & \\ & & & \vdots & & \ddots & \\ & & & u_n & & \cdots & 1 \end{pmatrix}$$

in which not all $u_1, \cdots, u_{j-1}, u_{j+1}, \cdots, u_n$ are zero.

5) Exist $T \in GL_n(K)$, such that $TAT^{-1} = T_{12}(1)$. Especially when $n \geq 3$ we can get $T \in SL_n(K)$.

6) Exist F is the subfield of K, such that A is similar to $I_{n-2} \oplus B$ in $GL_n(K)$, $B \in SL_2(F)$, $B \neq \pm I_2$, $\text{tr}B = 2$

Proof

1) $\Rightarrow$ 2) From the transvection definition $\exists P \in GL_n(K)$, such that $PAP^{-1} = T_{ij}(\lambda)$, so $\text{rank}(A - I) = \text{rank}(P(A - I)P^{-1}) = 1$ and from $P(A - I)P^{-1}$ is nilpotent matrix we have $(A - I)$ is nilpotent matrix, too.

2) $\Rightarrow$ 3) Exist P_1, $Q_1 \in GL_n(K)$, such that

$$P_1(A - I_n)Q_1 = E_{11}$$

order $P = P_{i1}P_1$, then $P \in GL_n(K)$, $P(A - I_n)Q_1 = E_{i1}$ we deduce

$$P(A - I_n)P^{-1} = E_{i1}Q_1^{-1}P^{-1} = \sum_{k=1}^{n} \lambda_k E_{ik}$$

through nilpotent matrix we have $\lambda_i = 0$, PAP^{-1} is the result form.

3) $\Rightarrow$ 4) $r(P(A - I)P^{-1}) = 1$ and $P(A - I)P^{-1}$ is nilpotent matrix, we deduce $A - I$ is nilpotent matrix either. So there exist P_2, $Q_2 \in GL_n(K)$, such that

$$P_2(A - I_n)Q_2 = E_{1j}$$

through $Q_2^{-1}(A - I_n)Q_2$ is nilpotent matrix $\Rightarrow Q_2^{-1}AQ_2$ is the result form.

4) $\Rightarrow$ 5) Get $Q \in GL_n(K)$

$$QAQ^{-1} = \begin{pmatrix} 1 & & & \\ \mu_2 & 1 & & \\ \vdots & & \ddots & \\ \mu_n & & \cdots & 1 \end{pmatrix}$$

in which not all $\mu_2, \cdots, \mu_n$ are zero. because there exist $N \in GL_{n-1}(K)$, such that

$$N \begin{pmatrix} \mu_2 \\ \vdots \\ \mu_n \end{pmatrix} = \begin{pmatrix} 1 \\ 0 \\ \vdots \\ 0 \end{pmatrix}$$

then

$$\left(\begin{pmatrix} 1 & \\ & N \end{pmatrix} Q \right) A \left(\begin{pmatrix} 1 & \\ & N \end{pmatrix} Q \right)^{-1} = T_{21}(1)$$

from

$$\begin{pmatrix} 0 & 1 \\ 1 & 0 \end{pmatrix} \begin{pmatrix} 1 & \\ 1 & 1 \end{pmatrix} \begin{pmatrix} 0 & 1 \\ 1 & 0 \end{pmatrix}^{-1} = \begin{pmatrix} 1 & 1 \\ & 1 \end{pmatrix}$$

we can deduce A is similar to $T_{12}(1)$.

5) $\Rightarrow$ 6) Get $B = \begin{pmatrix} 1 & 1 \\ & 1 \end{pmatrix}$, from

$$\begin{pmatrix} 0 & I_{n-2} \\ I_2 & 0 \end{pmatrix} (B \quad 0) \begin{pmatrix} 0 & I_{n-2} \\ I_2 & 0 \end{pmatrix}^{-1} = \begin{pmatrix} I_{n-2} & 0 \\ 0 & B \end{pmatrix}$$

we can deduce the result.

6) $\Rightarrow$ 1) Because there exist $T \in SL_2(K)$

$$TBT^{-1} = \begin{pmatrix} 0 & 1 \\ -1 & trB \end{pmatrix} = \begin{pmatrix} 0 & 1 \\ -1 & 2 \end{pmatrix}$$

from

$$\begin{pmatrix} 1 & \\ 1 & 1 \end{pmatrix} \begin{pmatrix} 1 & 1 \\ & 1 \end{pmatrix} \begin{pmatrix} 1 & \\ 1 & 1 \end{pmatrix}^{-1} = \begin{pmatrix} 0 & 1 \\ -1 & 2 \end{pmatrix}$$

we can deduce A is similar to $T_{n-1,n}(1)$. So A is a transvection. The end.

Be similar to , we have propositions below .

Definition Let $A \in GL_n(K), n \geq 2$, if A is similar to

$$I_{n-1} \oplus a, \quad a \neq 1$$

then we call A is a dilatation. If $A \in SL_n(K)$ is a dilatation, then we call A is a S - dilatation.

If exist $m \in Z^+, m \geq 2, c \in C(K^*), c \neq 1$, such that A is similar to $I_{n-m} \oplus cI_m$, then we call A is a big dilatation.

Proposition 8 Let $A \in GL_n(K)$, $n \geq 2$, then A is a dilatation $\Leftrightarrow$ res$A$ = 1 and the residue matrix of A is not a unit matrix.

Words and Expressions

nilpotent adj. 幂零的
nilpotent matrix 幂零阵
dilatation n. 伸缩
S-dilatation S 伸缩
big dilatation 大伸缩

7.3 QUASICENTER MATRIX

Definition Let $A \in GL_n(K)$, $n \geq 2$

1) When $n = 2$, if $A \in C(GL_2(K))$, then we call A is a quasicenter matrix, too;

2) When $n = 3$, if $\exists\ c \in C(K^*)$, A is similar to
$$\begin{pmatrix} a & & \\ b & c & \\ & & c \end{pmatrix}$$
or $c \bar{\in} C(K^*)$, A is similar to
$$\begin{pmatrix} a & & \\ 1 & c & \\ & & c \end{pmatrix}$$
in which $a + c$, $ac \in C(K^*)$ then we call A is a quasicenter matrix;

3) When $n \geq 4$, if exist $c \in C(K^*)$, such that A is similar to the matrix below
$$\begin{pmatrix} a & & \\ b & c & \\ & & cI_{n-2} \end{pmatrix}$$
then we call A is a quasicenter matrix. Quasicenter matrix is also called quasicentral.

Lemma 1 Let $A \in GL_3(K)$, then A is similar to
$$\begin{pmatrix} 0 & 0 & * \\ 1 & 0 & * \\ 0 & 1 & * \end{pmatrix}$$
$\Longleftrightarrow$ exist $X \in K^3$, such that
$$X,\ AX,\ A^2 X$$
are linear independent.

Proof If there exist $P \in GL_3(K)$
$$P^{-1}AP = \begin{pmatrix} 0 & 0 & * \\ 1 & 0 & * \\ 0 & 1 & * \end{pmatrix}$$

let
$$P = (P_1, P_2, P_3)$$
is column vector form, then we have
$$AP_1 = P_2, \ AP_2 = P_3$$
so P_1, AP_1, A^2P_1 are linear independent. On the contrary, there exist $X_1 \in K^3$, X_1, AX_1, A^2X_1 are linear independent, let $Q = (X_1, AX_1, A^2X_1)$ then
$$Q^{-1}AQ = Q^{-1}A(X_1, AX_1, A^2X_1) = Q^{-1}(AX_1, A^2X_1, A^3X_1) = \begin{pmatrix} 0 & 0 & * \\ 1 & 0 & * \\ 0 & 1 & * \end{pmatrix}$$

The end.

Corollary Let $A \in GL_3(K)$, then A is not similar to
$$\begin{pmatrix} 0 & 0 & * \\ 1 & 0 & * \\ 0 & 1 & * \end{pmatrix}$$

$\Longleftrightarrow$
$$A^2 X \in L(X, AX), \forall X \in K^3$$
in which
$$L(X, AX) = \{Xk_1 + AXk_2 \mid k_1, k_2 \in K\}$$

Proof Through Lemma 1 the sufficiency is obvious. On the contrary, if A is not similar to $\begin{pmatrix} 0 & 0 & * \\ 1 & 0 & * \\ 0 & 1 & * \end{pmatrix}$ through Lemma 1, then
$$X, AX, A^2X$$
are linear independence, $\forall X \in K^3$. So not all $k_1, k_2, k_3 \in K$ are zero, such that
$$Xk_1 + AXk_2 + A^2Xk_3 = 0$$
when $X \neq 0$, if $k_3 \neq 0$, $A^2X = X(-k_1 k_3^{-1}) + AX(-k_2 k_3^{-1}) \in L(X, AX)$, if $k_3 = 0$, $k_1, k_2 \in K^*$, thus $AX = Xk$, $k = -k_1 k_2^{-1}$, we

have
$$A^2X = AXk \in L(X, AX)$$
The end.

Lemma 2 Let $A \in GL_3(K)$, $A \notin C(GL_3(K))$, then A is not similar to $\begin{pmatrix} 0 & 0 & * \\ 1 & 0 & * \\ 0 & 1 & * \end{pmatrix} \Longleftrightarrow A$ is similar to

$$\begin{pmatrix} 0 & a & 0 \\ 1 & b & 0 \\ 0 & 0 & c \end{pmatrix}$$

in which $c^2 = a + cb$, $c \in C(K^*)$ or $a, b \in C(K^*)$.

Theorem 1 Let $A \in GL_3(K)$, then A is not similar to the matrix below

$$\begin{pmatrix} 0 & 0 & * \\ 1 & 0 & * \\ 0 & 1 & * \end{pmatrix}$$

if and only if A is a quasicenter matrix.

Proof When A is a quasicenter matrix, Obviously A is not similar to $\begin{pmatrix} 0 & 0 & * \\ 1 & 0 & * \\ 0 & 1 & * \end{pmatrix}$ when A is similar to

$$\begin{pmatrix} a & & \\ b & c & \\ & & c \end{pmatrix} \quad c \in C(K^*)$$

suppose $b \neq 0$, from

$$\begin{pmatrix} 1 & -ab^{-1} \\ & b^{-1} \end{pmatrix} \begin{pmatrix} a & \\ b & c \end{pmatrix} \begin{pmatrix} 1 & -ab^{-1} \\ & b^{-1} \end{pmatrix}^{-1} = \begin{pmatrix} 0 & -ac \\ 1 & a+c \end{pmatrix}$$

and $c^2 = -ac + c(a + c)$, from Lemma 2 we can deduce that A is not similar to

$$\begin{pmatrix} 0 & 0 & * \\ 1 & 0 & * \\ 0 & 1 & * \end{pmatrix}$$

In the same way, for A is similar to

$$\begin{pmatrix} a & & \\ b & c & \\ & & c \end{pmatrix} \quad ac, a+c \in C(K)$$

we can deduce A is fitted the result.

On the contrary, we can prove it from Lemma 2.

Theorem 2   Let $A \in GL_n(K)$, $n \geq 4$, exist $P \in GL_n(K)$, $PAP^{-1} = (b_{ij})$, $\begin{pmatrix} b_{31} & b_{32} \\ b_{41} & b_{42} \end{pmatrix} \in GL_2(K)$ if and only if A is not a quasicenter matrix.

Theorem 3   Let $A \in GL_n(K)$, $n \geq 2$, then A is not a quaisicenter matrix if and only if

1) For the case $n = 2$, A is similar to $\begin{pmatrix} 0 & * \\ -1 & * \end{pmatrix}$

2) for the case $n = 3$, A is similar to $\begin{pmatrix} 0 & 0 & * \\ -1 & 0 & * \\ 0 & -1 & * \end{pmatrix}$

3) For the case $n \geq 4$, A is similar to $\begin{pmatrix} 0_2 & \\ -I_2 & * \\ 0 & \end{pmatrix}$, in which zero matrix $0_2 \in K_2$.

Words and Expressions

quasicenter *adj.* 拟中心的
quasicenter matrix 拟中心矩阵
linear independence 线性无关
column vector 列向量

Vocabulary Index

A

addition *n.* 加法
administrator *n.* 管理者,治理者,行政官
advent *n.* 到来,来临
albeit *conj.* 虽然
algebra *n.* 代数,代数学
algebra steps 代数步骤
algebraic *adj.* 代数的
algebraic relationship 代数关系
algorithm *n.* 算法,计算程序
algorithmic *adj.* 算法的,规则系统的
algorithmic process 算法过程
aloof *adv.* 远离,躲开 *adj.* 冷漠的,疏远的,无情的
alternator *n.* 交流发电机
AM arithmetic mean 算术平均值
analysis *n.* 分析,分析学;解析
analytical *adj.* 解析的,分析的
analytical geometry 解析几何
anarchy *n.* 无政府状态,混乱
angle *n.* 角
annuity *n.* 年金,养老金;年金保险
any point on a curve 曲线上任意一点
apoplexy *n.* 脑溢血
appreciable *adj.* 可见到的,可觉得的,可估计的

approach n. 近似值;近似法;方法,途径 vt. 趋近,接近
appropriate adj. 适当的,合适的,正当的
arbitrage n. 裁决,仲裁;(股票等)套利,套汇
arbitrary n. 任意的,随意的
arbitrary constant 任意常数
arc length 弧长
arccosine n. 反余弦
archaic adj. 古的,古代的;已废的,已不通用的
arcsecant n. 反正割
arctan = arc tangent 反正切
arctangent n. 反正切
area n. 面积
area – function 面积函数
array n. 排列;整列,排阵;序列 vt. 排列,布置;装饰
ascending adj. 上升的;向上的;增长的
ascending power 升幂
asset n. 有价值的特性或技能,(常用复数)财产,资产
assign vt. 指定;分配
assigning value 指定值
associative adj. 结合的
asymmetry n. 非对称
augment vt. or vi. 增大,增加
augmented matrix 增广矩阵
axis n. 轴
axiom n. 公理,公设

B

base n. 底,基
basis n. 基
belch vt. 吐气,吐出 vi. 打嗝,喷出 n. 大嗝,嗝;喷出之物

binomial *n*. 二项式 *adj*. 二项的,二项式的

the binomial theorem 二项式定理

blurt *vt*. 不假思索地脱口说出 *n*. 不假思索的话

bog *n*. 沼泽 *vt*. 使陷于泥潭

bond *n*. 有息债券;合同;票据

bribery *n*. 行贿或受贿的行为

bummer *n*. [美俚]依赖他人而过活的人,懒虫,饭桶

bumper *n*. 满杯; *vt*. 倒满杯

bureaucrat *n*. 官僚

C

cancel *vt*. 消去,消除;*vi*. 抵消 *n*. 约分

calculus *n*. 微积分,微积分学;演算

catastrophic *adj*. 悲惨的,毁灭的

caterpillar *n*. 蝴蝶或飞蛾类之幼虫;毛虫

cease *vi*. 停止;终止

chance *n*. 机会

characteristic *adj*. 特征的,示性的

characteristic equation 特征方程

clarity *n*. 明晰

coefficient *n*. 系数

coincide *vi*. 一致,符合;空间上相合;时间上相合

coloration *n*. 染色;着色;色泽

column vector 列向量

commutative *adj*. 交换的

commutator *n*. 换位子

commutator group 换位子群

commutator of transvection 平延换位子

compass *n*. 圆规,指南针,范围

complex *n*. 复形 *adj*. 复的

complex number 复数
component *adj*. 组成的,成分的 *n*. 成分,分力
conceptualization *n*. 概念化
conjugate *adj*. 共轭的
conjugate element 共轭元
conscious *adj*. 觉得的,知道的;有意识的
consistent *adj*. 相容的,一致的,无矛盾的
consistent method 一致的方法
conspicuous *adj*. 显著的,显而易见的,引人注目的
constant *n*. 常项,常数
constant term 常数项
construct *vt*. 组成,构造,设计;作(圆)
content *n*. 知足,满足,满意,愉快
continuous *adj*. 连续的
continuous distribution (or **cumulative distribution function**)连续分布
continuous function 连续函数
contribute *vt*. 贡献,捐助
convex *n*. 凸 *adj*. 凸的
convex function 凸函数
convexity *n*. 凸性
core *n*. 核心,柱心
correspond *vi*. 对应
corresponding *adj*. 同位的,对应的(与 *to* 连用)
coupon *n*. 证明持有人有做某事或获得某物之权利的票据
criterion *n*. *pl*. criteria 准则,判别准则,判据
cryptic *adj*. 秘密的,隐藏的,潜隐的
cryptic coloration 保护色
cryptically *adv*. 隐性地,隐蔽地,秘密地
cube *n*. 立方,立方体
cumulative *adj*. 累积的

curve *n*. 曲线
cycloid *n*. 摆线,旋轮线

D

decent *adj*. 适当的,可接受的;令人满意的,得体的
decomposition *n*. 分解
definite *adj*. 定的,确定的
demonstration *n*. 证明,示教
denominator *n*. 分母
derivative *n*. 导数,微商 *adj*. 非独创的,由他事物演变的
derive *vt*. 获得,得来;起源,由来 *vi*. 起源,发生
derived from 起源,发生
despise *vt*. 轻视,蔑视
determinant *n*. 行列式
diagnosis *n*. 诊断,判断问题,确定毛病;诊断结果,诊断书
diagonal *adj*. 对角的,对角线的
diagram *n*. 图,图表,图解
die *n*. 骰子 pl. *dice* ;小立方体;印膜
difference *n*. 差,差分
differentiable *adj*. 可微的
differentiable function 微分方程
differential *n*. 微分,微分的
differential equation 微分方程
twice differentiable function 二阶微分方程
differentiation *n*. 微分法
diffusion *n*. 散步,传播,弥漫,扩散
dilatation *n*. 伸缩
S-dilatation S 伸缩
big dilatation 大伸缩
dimension *n*. 维,维数;量纲

three dimensions 三维
diminish *vt*. 减少,缩小,使变小
directrix *n*. 准线
disprove *vt*. 反证,证明为误
distinct *adj*. 分开的,不同的
distribution *n*. 分布,分配;
distributive *n*. 分配的
distributive law 分配率
dividend *n*. 股息,红利
division *n*. 除法,除
dominance *n*. 统治,支配
dominant *vt*. 统治,支配 *vi*. 管辖,支配 *adj*. 支配的
dominant eigenpair 主特征对,主本征对

E

eccentric *adj*. 古怪的,怪癖的;不同中心的(圆);离中心的,偏心的
egalitarian *adj*. 相信人人平等的 *n*. 平等主义
eigenpair *n*. 特征对,本征对
eigensystem *n*. 本征系统
eigenvalue *n*. 特征值
eigenvector *n*. 特征向量
element *n*. 元,元素
elimination *n*. 消元法,消去
ellipse *n*. 椭圆
empty *adj*. 空
empty set 空集
entry *n*. 表值;进入,通路
equal *vt*. 等于,相等 *adj*. 相等的
equality *n*. 等式,相等
equally *adv*. 相等地,同样地

equally likely 等可能的
equation $n.$ 方程
the Diophantine equation 丢番图方程
equilibrium $n.$ 平均,均衡,均势
equilibrium value 平衡值
equivalently $adv.$ 等价地,等势地,相等地
essence $n.$ 本质,精髓,要素
essential $n.$ 必要的;本质的
estate $n.$ 地产,财产
estimate $vt.$ 估计 $n.$ 估计量
even $adj.$ 偶的,偶数的;整数的
even number 偶数
event $n.$ 事件
exhaustion $n.$ 穷举,耗尽
experiment $n.$ 实验,试验
expiry $n.$ 终止,届满,期满,满期
exploration $n.$ 探险,仔细检查
exponent $n.$ 指数,幂
exponential $n.$ 指数的
exponential function 指数函数
extension $n.$ 扩张,开拓;外延
extinct $adj.$ 灭种的,灭绝的;熄灭了的
extirpate $vt.$ 根除,灭绝,连根拔起
extrema $n.$ ($pl.$ form of extremum) 极值

F

factor $n.$ 因数,因子,遗传因子
fervour $n.$ 热诚,热心;热,炙热
fiddle $vi.$ 做无益之事,乱动 $vt.$ 演奏,虚度(光阴)
field $n.$ 域

figure $n.$ 图,图形;数字
first order differential equation 一阶微分方程
fit $vt.$ 安装,装备;使适应
fitting $n.$ 拟合
fix space 固定空间
flaws $vt.$ 使有裂痕,使有瑕疵 $vi.$ 有裂痕,有瑕疵 $n.$ 裂缝,裂纹;缺陷,瑕疵
fluxions $n.$ 流数
focus $n.$ 焦点
formula $n.$ 公式
frequency $n.$ 频率,时常发生
friction $n.$ 摩擦,人或党派之间不同观点的矛盾或冲突
function $n.$ 函数;函词,函项
fundamental $n.$ 基本的

G

gadget $n.$ [俗]设计精巧的小机器
Gaussian 高斯
Gaussian Elimination 高斯消元法
geneticist $n.$ 遗传学者
genetics $n.$ 遗传学
general linear group 一般线性群
general solution 通解
generality $adj.$ 普遍性,普通性,一般性
without loss of generality 不失一般性
genotype $n.$ 基因型,遗传型
geometric mean 几何平均
geometrical $adj.$ = $geometric$ 几何的,几何学的
geometrical meaning 几何含义
geometry $n.$ 几何,几何学

gobble *vt. or vi.* 大吃,狼吞虎咽
graph *n.* 图,图形;网络
grave *n.* 坟墓,墓碑;死地,死 *adj.* 庄重的,严肃的,重大的;阴沉的
Gresham College 葛莱兴学院
grime *n.* 污秽 *vt.* 使污浊,使覆有污秽物
group *n.* 群
guarantee *n.* 担保,保证;担保人,保证人;保证书 *vt.* 保证,担保

H

hanker *vi.* 渴望,眷恋
harmonic *n.* 调和,调和函数 *adj.* 调和的
harmonic mean 调和平均,调和中项,调和中数
harsh *adj.* 粗糙的;严厉的,苛刻的;恶劣的
hedge *n.* 防止造成损失的手段;树篱
height *n.* 高,高度
homogenous *n.* 齐次的,齐的;同类的
linear homogenous equation 线性齐次方程
homozygous *adj.* 同型结合的,纯合子的
hone *vi.* 发怨声,悲叹 *n.* 细磨刀石
hyperboloid *n.* 双曲面,双曲线体
hyperboloid of revolution 旋转双曲面
hypothesis *n.* 假设

I

idealize *vt.* 理想化
identical *adj.* 恒等的,恒同的
identity *n.* 恒等,恒等式,单位元
identity matrix 单位矩阵
if and only if 当且仅当
ignition *n.* 点火,燃烧,点火装置

imaginary *adj.* 虚的
impending *adj.* 即将发生的,迫在眉睫的,行将到来的
impenetrable *adj.* 不可理解的,难以探究的
improvisation *n.* 即席创作;即兴
incur *vt.* 遭受,蒙受,招致
indefinite *adj.* 不定的
index *n.* 指标;指数;下标;索引 *vt.* 标引
indicate *vt.* 指示,指出;显示;表示
indivisible *adj.* 除不尽的
induction *n.* 归纳,归纳法
induction hypothesis 归纳假设
indulge *vt.* 放任,纵容;迁就 *vi.* 纵情,任意
inequality *n.* 不等式,不等
infection rate 感染率
infinite series 无穷级数
infinitesimal *n.* 无穷小,无限小 *adj.* 无穷小的,无限小的
infinity *n.* 无穷大,无穷
informative *adj.* 有益的,供给知识的
inhabitant *n.* (某地的)居民,住户,栖息的动物
inherent *adj.* 内在的,固有的,本质的
inherit *vt.* 继承,由遗传而得 *vi.* 继承,接受遗传的力量、特点等
informative *adj.* 有益的,供给知识的,
innumeracy *n.* 无数学基本功,不识数
integer *n.* 整数
integer solution 积分解
integral *n.* 积分,整数 *adj.* 积分的,正的
integral calculus 积分学
integration *n.* 积分,积分法
intermediary *n.* 中间人,调节人
intermediate *adj.* 中间的,中级的,中间人的 *n.* 中间物,中间人,调停者

intersect *vt*. 交,相交
intersections *n*. 交,相交
interval *n*. 区间
invert *vt*. 上下颠倒,倒转,转回
inverse *n*. 逆
inverse element 逆元
inverse function 反函数
inverse tangent 反正切
invert *vt*. 上下颠倒,倒转,转回
invertible *n*. 可逆的
invertible matrix 可逆矩阵
Israeli *n*. 以色列;犹太人

K

keen *adj*. 热心的,渴望的;锋利的,尖锐的
keener *n*. 热心者

L

land *n*. 陆地,国土 *vt*. 使降落,使着陆 *vi*. 登岸,着陆
land on 落在(一个较小的)物体上
least *adj*. 最小的,最少的 *adv*. 最小,最少
least square fitting 最小二乘方拟合
least upper bound 上确界
leather *n*. 皮革;皮革制品 *adj*. 皮革制的,皮的
leather merchant 皮革商人
lethal *adj*. 致命的,致死的,毁灭性的
lichen *n*. 青苔,地衣,苔藓
like term 同类项
likelihood *n*. 似然
likely *n*. 可能的

line segment 线段
linear *adj.* 线性的,一次的
linear algebra 线性代数
linear equation 线性方程
linear function 线性函数,一次函数
linear group 线性群
linear independence 线性无关
Linear Systems of Equations 线性方程组
lofty *adj.* 高的,高尚的;高傲的
longitude *n.* 经度
Lutheran *n.* 路德教徒 *adj.* 马丁路德的,路德教派的

M

manifold *adj.* 多重的,繁多的;多倍的,同时做多种事 *n.* 簇,流形;复写本;复写纸;多重
manipulation *n.* (用手的)操纵;巧妙的操纵
manuscript *n.* 手稿,草稿,原稿
marvelous *adj.* 奇异的,不平常的,了不起的
mathematical *adj.* 数学的
mathematical analysis 数学分析
mathematical induction 数学归纳法
matrix *n. pl. matrices or matrixes* 矩阵,真值表
matrix algebra 矩阵代数
matrix addition 矩阵加法
matrix equation 矩阵方程
matrix inverse 矩阵的逆
maxima *n.* (*pl. form of maximum*) 极大值,极大
maximize *n.* 使最大
meaning *n.* 意义,含义
melantic *adj.* 黑变病的,黑性的,黑色素过多的

member *n*. 元
members of the set 集合的元
mental arithmetic 心算
method *n*. 法,方法
method of exhaustion 穷举法
method of indivisibles 不可分量法
meticulous *adj*. 极端注意琐事的,拘泥于细节的
mild *adj*. 温柔的,和善的,温暖的,宽大的,不严重的
mimic *vt*. 仿效,拟态;与…极相似 *n*. 擅模仿的人或物
mimicry *n*. 拟态,模仿
minima *n*. (*pl*. form of minimum) 极小值
minimal *adj*. 极小的
ministerial *adj*. 牧师、内阁或部长的,部长级的;行政上的;附属的
mire *n*. 泥泞,沼池,泥浆,困境 *vt*. 使陷入泥泞,使陷入困境
moth *n*. 虫,蛾
multiples *n*. 倍数,倍式;多重
multiplication *n*. 乘法
multitude *n*. 众多,群众

N

negation *n*. 否定,否定词
negative *adj*. 否定的;负的
neophyte *n*. 初学者
Netherlands *n*. 荷兰
nilpotent *adj*. 幂零的
nilpotent matrix 幂零阵
non-invertible 非可逆的
non-square matrix 非方阵
non-zero 非零
non-zero real number 非零实数

non-zero solution 非零解
notation $n.$ 记号,记法
notorious $adj.$ 声名狼藉的,著名的,众人皆知的
numerator $n.$ 分子

O

obsolescence $n.$ 将要过时
obsolete $adj.$ 作废的,已废的;过时的
odd $adj.$ 奇数的
odd number 奇数
operation $n.$ 运算
operator $n.$ 算子
optics $n.$ 光学
option $n.$ 可供选择的事物,选择
ordain $vi.$ 注定,任命 $vt.$ 注定,规定;任命
order $n.$ 序,次;阶,级 $vt.$ 定货
ordered set 有序集
ordinary $adj.$ 常的,寻常的,正常的
ordinary differential equation 常微分方程
outcome $n.$ 结果

P

parabola $n.$ 抛物线
paraboloid $n.$ 抛物面
paraboloid of revolution 回转抛物面
parallel $adj.$ 平行的,并行的 $n.$ 平行线
parallel lines 平行线
parameter $n.$ 参数,参变量
partial $adj.$ 偏的;部分的
partial fraction 部分分式;部分分数

partial differential equation 偏微分方程
participant $n.$ 参加者
participate $vi.$ 参加,参与
particular solution 特解
perpendicular $n.$ 垂线 $adj.$ 垂直的,垂直于
perpendicular to the y – axis 垂直于 y 轴
per unit length 每个长度单位
phenotype $n.$ 显型,表型;显型类之所有生物
phenotypically $adv.$ 显型地
plane $n.$ 平面
plumber $n.$ 铅管工人
point $n.$ 点
a moving point 一个动点
polynomial $n.$ 多项式
periodic $adj.$ 定期的,周期的
portfolio $n.$ 投资组合;公文包,文件夹
positive $adj.$ 正的,肯定的
positive integer 正整数
possible $adj.$ 可能的
possible outcome 可能的结果
power $n.$ 幂,乘方;势;权
power mean 幂平均
power mean inequality 幂平均不等式
Prague 布拉格(捷克首都)
precedes $vt.$ 在前,在先,优于 $vi.$ 在前,在先
precisely $adv.$ 精确地;严格地
precision $n.$ 精度,精确度,精确
predatory $adj.$ 抢夺的,有抢夺性的
predominantly $adv.$ 主要地,杰出地
premium $n.$ 保险费,额外费用,津贴,花红;

priority $n.$ 优先权
probability $n.$ 概率
probability function（or **probability distribution function**）概率函数
procedural $adj.$ 程序的,程序上的,有关程序的
process $n.$ 方法,步骤;过程
product $n.$ 积,乘积
product of eigenvalues 特征值的积
proper $adj.$ 真,正常,常态,固有的
proper subset 真子集
property $n.$ 性质
proportional $adj.$ 成比例的,相称的（习语）(指公债或股票)超过正常或市面的价值,溢价
Pythagorean $n.$ 毕达哥拉斯学派 $adj.$ 毕达哥拉斯的

Q

quantitative $adj.$ 数量的,有关数量的
quantitatively $adv.$ 与量有关地,定量地
quasicenter $adj.$ 拟中心的
quasicenter matrix 拟中心矩阵
quasi-diagonal 拟对角的
quasi-diagonal matrix 拟对角矩阵
quasidilatation $n.$ 拟伸缩
quasistability $n.$ 准稳定的,类似稳定的
quiz $n.$ [美]非正式的考试
quotient $n.$ 商

R

ramble $v.$ 漫步,闲逛;漫谈,闲聊
random $adj.$ 随机的
random variables 随机变量

range *n*. 值域,范围,变程

real *adj*. 实的,实数的,有效的 *n*. 实数,实部,实型

real-valued function 实值函数

reason *n*. 原因,前提;理由,推理 *vt*. 说服,劝说

reasoning *n*. 推理,推论,推导,推断

recessive *adj*. 隐性的,退后的

reciprocal *n*. 倒数 *adj*. 倒数的,互反的,互逆的

reckon *vt*. 计算,估算 *vi*. 计算,算账

reckoning *n*. 计算,统计,核算

recolonize *vt*. 再殖民于,重新殖民于

rectangle *n*. 矩形(又称"长方形")

reducible *n*. 可归约的,可简化的

region *n*. 区域

relation *n*. 关系

relationship *n*. 关系

relent *vi*. 变温和,变宽容 *vt*. 使变温和,使变宽容

relentless *adj*. 无慈悲心的,不放松的

remark *n*. 注意,意见,附注(常用复数)

remedial *adj*. 修补的,矫正的;补救的;治疗的

reputable *adj*. 声誉好的,有名望的,值得信赖的

residue *n*. 剩余,留数

residue matrix 剩余矩阵

residue number 剩余数

residue space 剩余空间

respectively *adv*. 各自地,个别地

restricted *adj*. 限制的,约束的

reveal *vt*. 泄露,透露;显示,显出

reversion *n*. 反回

revolution *n*. 回转

rigid *adj*. 僵硬的,稳固的;严格的,严厉的;精确的

right vection space 右向量空间
ring $n.$ 环
robust $n.$ 鲁棒 $adj.$ 稳健
root $n.$ 根
root mean square 均方根
rote $n.$ 机械的方法,固定程序;背诵,强记
routine $n.$ 例行公事,惯例,常规
row $n.$ 行 ith row 第 i 行

S

salute $n.$ 致意,问候;敬礼 $vt.$ 向…敬礼,向…致意 $vi.$ 致意,祝贺
scalar $n.$ 标量,数量
scalar multiplication 数量乘法
eccentric $adj.$ 古怪的,怪癖的;不同中心的(圆);离中心的,偏心的
secant $n.$ 正割
security $n.$ 安全,保护,保障;安全措施;抵押品
segment $n.$ 段,节
a line segment of length a 一个长度为 a 的线段
sequence $n.$ 序列
series $n.$ 级数,列
set $n.$ 集,集合
shaded regions 阴影区域
significantly $adv.$ 值得注目地,非偶然地;另有含义地
significantly contributed 值得注目地贡献
similar $adj.$ 相似的
simplistic $adj.$ 过分简单的
skewfield $n.$ 体
slide $vi.$ 滑动;滑行,溜过 $vt.$ 使滑动,使轻轻溜过
slope $n.$ 斜率
snap $vt.$ 咬,攫取

solemnity *n*. 庄严,严肃

solution *n*. 解,解法

solution space 解空间

soot *n*. 煤烟,煤灰,油烟 *vt*. 使某物蒙上黑烟灰

sophisticated *adj*. 复杂的,完善的,成熟的

special linear group 特殊线性群

specify *vt*. or *vi*. 指定,载明

spectrum *n*. 范围,系列;光谱

speculation *n*. 思考,思索,推断,推测

sphere *n*. 球面,球形

spin *vt*. 自转,旋转

spinner *n*. 旋转物

spiral *n*. 螺线,蜷线

square matrix 方阵

statistics *n*. 统计学;统计,统计表

step *n*. 阶段,级,步骤,步长

stochastic *adj*. 猜测的,好猜测的

stock *n*. 股份(常用复数);公债,存货

strategy *n*. 战略;策略,谋略;对策,政策

strictly *adv*. 严格地

strictly convex 严格凸函数

strictly increasing function 严格增函数

strident *adj*. 作粗糙声的,发尖锐声的

subconscious *adj*. 潜意识的 *n*. 潜意识,下意识

subfield *n*. 子域

subset *n*. 子集

substitute *vt*. 代换,代入 *n*. 代替者,代理人;代替物,替代品

subtract *vt*. 减去

such that 使得

sum *n*. 和,总数

summon $vt.$ 召唤,传唤;召集
surface $n.$ 面,曲面
surface area of a sphere 球的表面
susceptibility $n.$ 感受性
symbolic $adj.$ 符号的
symbolize $vt.$ 以符号表示
symmetric $adj.$ 对称的
systematic $adj.$ 系统的,有系统的
systematic method 系统法

T

Taylor series 泰勒级数
tangent $n.$ 正切,切线 $adj.$ 正切的,切线的
tangent lines 切线
term $n.$ 术语,项,条
the definite integral 定积分
the focus of the parabola 抛物线的焦点
the Fundamental Theorem of Algebra 代数基本定理
the indefinite integral 不定积分
the natural logarithm function 自然对数
the product of the base and height 底与高之积
the St. Petersburg Academy 圣比德堡科学院
the tangent to a curve 曲线的切线
theorem $n.$ 定理
the Fundamental Theorem of Calculus 微积分学基本定理
theory of numbers 数论
total number 总数
total number of possible outcome 所有可能的结果
transform $vt.$ 使变形
transvection $n.$ 平延

simple transvection 初等平延
trial $n.$ 试,试验
triangle $n.$ 三角,三角形
triangle inequality 三角不等式
trigonometric $adj.$ 三角的,三角学的,三角法的
trigonometric functions 三角函数,圆函数
triply $adv.$ 三重地,三倍地
trivially $adv.$ 平凡地
tuberculosis $n.$ ($abbr.$ TB)结核病,(尤指)肺结核
tutor $n.$ (英国大学)导师;(美国大学)教员(职位低于讲师)
tutorial $adj.$ 家庭教师的,(大学)导师的
tyranny $n.$ 虐政;暴行

U

ultimately $adv.$ 最后;最终;终结
underdetermined $adj.$ 欠定,亚定
underneath $prep.$ 在…之下
unions $n.$ 并,并集
unique $adj.$ 惟一的
unique solution 惟一解
unit circle 单位圆
universal theory of gravitation 万有引力
upper triangular form 上三角形

V

vacuum $n.$ 真空;真空吸尘器 $adj.$ 真空的;产生真空的
vaguely $adv.$ 不明确地,不清楚地;无表情地
value $n.$ 值,数值
variable $n.$ 变量,元,变元;变项
vector space 向量空间

ventured $n.$ 工作项目或事业,商业,企业 $v.$ 敢于去,敢说
versed sine 反正弦
version $n.$ 译文,翻译
versus $prep.$ 对;相形,比较
visualization $n.$ 想像
volatile $adj.$ 不稳定的,无常性的
volatility $n.$ 易变,轻快;挥发,挥发性
voltage $n.$ 伏特数,电压,电压量
volume $n.$ 体积,容积
vulnerability $n.$ 弱点,容易受伤害
vulnerable $adj.$ 可伤害的,易受攻击的;难防守的,有弱点的

W

weighted $adj.$ 加权的
weighted AM – GM Inequality 加权算术平均几何平均不等式
weighted arithmetic mean（**WAM**）加权算术平均
weighted geometric mean（**WGM**）加权几何平均
weighted harmonic mean（**WHM**）加权调和平均
weighted mean 加权平均
weighted root mean square（**WRMS**）加权均方根

X

x-axis x 轴

Y

y-axis y 轴
yearn $vi.$ 渴望,思念;怀念,怜悯
Yugoslavia $n.$ 南斯拉夫